CW00498093

The Construction of the Modern Locomotive

LIBRARY - Name & Address

[illegible]

TJD/H87 2966 6143

TRIM SIZE	1 2 3 4 5 6 7 8 9 10
Buckram Color # 192	Your Spine will be lettered **EXACTLY** as it appears on your Binding Slip.
Stamp in White ☑	CONSTRUCTION OF THE MODERN LOCOMOTIVE HUGHES
Stamp in Black ☐	
Stamp in Gold ☐	
NEW: ☐	
Bound Before ☑	
Rub Enclosed ☐	
Sample ☐	
ADS:	
Leave In ☐	
Take Out ☐	
COVERS:	
Remove ☐	
Bind in All ☐	
Bind in Front ☐	
☐	+1 S
INDEXES:	F-F
Front ☐ Stub For ☐	
Back ☐ No Index ☐	

30 29 28 27 26 25 24 23 22 21 20 19 18 17 16 15 14 13 12 11 10 9 8 7 6 5 4

SPECIAL INSTRUCTIONS: [illegible]

Bound By AMERICAN BINDERY, Topeka, Kansas.

BINDERY COPY

R. JOHN A. F. ASPINALL, M. INST. C.E.,

THE CONSTRUCTION

OF THE

MODERN LOCOMOTIVE

BY

GEORGE HUGHES

ASSISTANT IN THE CHIEF MECHANICAL ENGINEER'S DEPARTMENT,
LANCASHIRE AND YORKSHIRE RAILWAY

London:
E. & F. N. SPON, 125 STRAND

New York:
SPON & CHAMBERLAIN, 12 CORTLANDT STREET

1894

Dedicated to

J. A. F. ASPINALL, Esq., M. Inst. C.E., Etc.,

CHIEF MECHANICAL ENGINEER TO

THE LANCASHIRE AND YORKSHIRE RAILWAY,

AS A CORDIAL AND GRATEFUL ACKNOWLEDGMENT OF

MUCH KINDNESS SHOWN TO THE AUTHOR

DURING THE PAST SEVEN YEARS.

PREFACE.

MANY books have been written upon Railways and Railway work, both from a non- and semi-professional point of view, which no doubt supply some particular want, but the author trusts that the present work may prove acceptable in a wider sense, not only to those who are engaged in responsible management and designing, but also to the thousands occupied in the workshops and on the footplate.

The work has been divided into sections, this proving the most satisfactory way, and where necessary, subdivided into parts, viz. Boiler Shop Practice; Foundry (Iron, Steel and Brass); Forge (Smithy, including Springs); Coppersmiths' work; Machine Shop; Erecting, &c. Design has not been touched upon, as it does not come within the scope of the author's plan. Each section describes, step by step, more minutely, and by such drawings and illustrations as have not appeared before in one volume, the actual progress of the work done in that section. There is one exception. The use of Steel Castings. To have described their manufacture would have entailed a lengthy dissertation upon the making of steel generally, with particular reference to the overcoming of such difficulties as oxidation, the formation of blow-holes, annealing, and the preparation of moulds; therefore, it was considered more useful to the general consumer, to devote a portion of a section to their use in conjunction with their chemical and physical properties, with a view to reduce to uniformity the various opinions held by individual consumers, and thus to formulate a standard specification.

Finally, it is my pleasing duty to express thanks to those who have, by word and deed, assisted me in gaining, or supplying information, and especially so to the Editor of 'The Engineer,' at whose suggestion this work was undertaken, the major portion of which has appeared in the pages of that journal.

GEO. HUGHES.

HORWICH, LANCASHIRE:
April 1894.

CONTENTS.

SECTION I.—THE BOILER.

SECTION II.—THE FOUNDRY.

PART I.—IRON FOUNDRY.

PART II.—THE USE OF STEEL CASTINGS.

PART III.—BRASS FOUNDRY.

SECTION III.—FORGINGS.

PART I.—FORGE.

PART II.—SMITHY, INCLUDING SPRINGS.

SECTION IV.—COPPERSMITHS' WORK.

SECTION V.—THE MACHINE SHOP.

SECTION VI.—ERECTING.

THE CONSTRUCTION

OF

THE MODERN LOCOMOTIVE.

SECTION I.

THE BOILER.

IN all well-regulated boiler shops, whether the plates are manufactured on the ground or bought from the steel makers, the tenacity of every plate used is known, and the number of that plate registered. Provision is also made for obtaining the analysis of the cast from which those plates were made if necessary. Generally the facilities offered by plate makers for testing and examining surfaces are adequate. The author has known the plates for 200 locomotive boilers to be delivered, and every plate worked up. The plates rejected during inspection for high tenacity, short elongation or bad surfaces were comparatively few.

The quality of the plates used in the boiler practice of to-day is that known as the "best mild steel," of an ultimate tenacity of not less than 26 tons per square inch nor more than 30 tons per square inch, with a minimum elongation of 20 per cent. on an 8-inch bar. Generally the tenacity of plates made by well-known firms ranges between 27 and 29 tons per square inch, and 25 to 32 per cent. elongation, although plates are known with a minimum of 24 tons per square inch with an elongation of 32 per cent.; the above results being upon test bars taken from the direction of the rolling. The tests from the opposite direction may, and often do, approach the lengthway for tenacity and elongation; but an average of fifty crossway tests, taken at random over a period of six months, gave 27·25 tons per square inch, and an elongation of 25 per cent. on 8 inches.

Regarding the chemical analysis, the plates must be free from silicon, sulphur and phosphorus. The analysis is generally found to be:—Carbon, about ·16 to ·18; silicon, ·01 to ·018; sulphur, ·03 to ·05; phosphorus, ·02 to ·04; and manganese, ·25 to ·48. All plates must be rolled truly to a uniform thickness, and both sides must be free from pitting, scale, dirt, lamination or other defects. After shearing, boiler plates must be annealed, and in no case must they leave the mill floor until all buckling is taken out, and they are sufficiently level and true for machining. A boiler shop is never absolutely free from this, lifting and transit always causing unevenness in plates, which is dealt with later on. The maker's name and test number must always be stamped upon the plates about 12 inches from each edge, at one corner, and to facilitate inspection this is encircled with a white paint mark. From the appearance of the surface and the edge of the plates it can be seen at a glance whether they have been annealed or not. Test strips, after being heated and cooled in water at about 80° F., should bend to a curve, the inner radius of which is one and a-half times the thickness of the plate. It is generally found that the majority will more than satisfy this test, bending down flat upon themselves. Drifting tests are made periodically upon strips 3 inches wide, in which $\frac{5}{8}$ inch holes are punched, and afterwards drifted cold until they reach $1\frac{1}{8}$-inch diameter. Some specifications stipulate that the plates are to be made from ingots hammered upon all sides, which gives a more homogeneous material, cogging being equally good.

The modern boiler shop is replete with first-class tools, and the template is used throughout, consequently the finest material is used to its best advantage, and the modern boiler is almost a perfect work of accuracy. Angle iron, &c., is nearly dispensed with, being replaced by hydraulic flanging; circular flanges are turned and the edges of all plates are planed. Almost all the rivets are closed by hydraulic pressure, and are of the best quality. It is generally admitted that rivet steel should be of the toughest material, so that test specimens should bend through an angle of 180°, and close down upon themselves when cold, without showing a crack or flaw. The tensile strength should be from 24 to 28 tons per square inch, with 25 per cent. elongation on 8 inches, and 45 per cent. contraction of area, and showing good tough silky structure. A tenacity of 26·5 to 27 tons per square inch is sufficient, because the mechanical work put into the material in making the rivet, will increase its tenacity in some cases from 2 to 4 per cent. The metal must further be capable of going through the various processes of manufacture without serious abrasion and cracking. It should flow well in riveting, and the heads of test samples should admit of being hammered down to ⅛ inch in thickness without fracturing the edges when hot. The carbon should not be more than ·15 per cent. and the phosphorus ·04, although some American specifications admit of phosphorus ·07.

Whenever the term "best iron" is used, it must be understood to be a well-known brand, and the qualification "best mild steel" infers an established firm; although, perhaps, lately the testing machine has done much to show up the good quality material made by small firms. Whenever the word "or" appears, it is to be distinctly understood that only one or the other item would appear in a specification.

A modern shop may be conveniently built in three bays,

the one containing the various drilling, planing, punching, shearing and bending machines, suitably situated in order to carry work forward as much as possible. The larger machines, at which heavy work is done, are provided with jib and lifting tackle. The second, or central division may be devoted to the marking of plates, the fitting together of the shells, and the general erection of boilers; whilst the third bay may take the general repairs. The angle smiths and hearths for welding internal flues, domes, &c., are placed in a smithy immediately adjoining the boiler shop. The hydraulic riveters attached to jibs of various rake, should be conveniently placed for the work they are required to do, as also the pumps and accumulators. The shops should also be fitted with the necessary pressure and return piping from the accumulator to the various hydraulic machines, whilst a narrow-gauge tramway should serve the machines and heavier portions of the work. Convenient pits should be made for certain machines, to facilitate drilling flanges and foundation rings, although the specially designed flange-drilling machine has rendered them less necessary. Pits are also required for certain punching machines, multiple drills, riveting, &c., and are convenient for fullering the foundation ring and fitting on the roof bars. Each bay should be traversed from end to end by a 12 or 15-ton rope-driven crane. The rivet fires may be placed between columns, &c., wherever required.

The workmanship is of a finished and accurate character all rivets completely filling the holes, which are slightly countersunk under the rivet heads, all holes in the plates and flanges being perfectly fair with each other, and no drifting is allowed, in the accepted sense of the term, on any consideration. If any of the holes are not perfectly fair with each other, they are rimered until they become so, care being taken that after rimering, the rivets shall completely fill the holes. All the plates are brought well together before any rivets are put in. Outside edges of holes are slightly countersunk and all burrs removed. The pitch of the rivets and the lap of the plates are shown in the detail drawings, Figs. 1–5, and rivets are countersunk only where mentioned on the drawings, see Fig. 13. The edges of all the plates are planed, turned, or shaped to an angle of one in eight before being put together, so as to leave a full edge for fullering, which is done with a broad-faced fuller actuated by pneumatic pressure, the ordinary caulking tool being dispensed with. It is quite unnecessary to go into the various rules and data used in the workshop by the workman in getting circumferences, &c., as these can readily be turned up in the various books published for the boiler-maker.

After the plates are delivered from the mill floor to the boiler yard, having been rolled and sheared in lengths and widths according to the position they will eventually occupy, they are at once taken in hand, and where necessary, the small amount of buckle that may remain in them, is taken out and otherwise straightened.

If a plate is buckled in the middle the edges have to be elongated, and if the edges are buckled the operation is *vice versâ*, a plater placing his flatter where his judgment indicates, and gradually elongating wherever required, which often occupies hours of tedious work. This process is greatly shortened by a multiple roller straightening machine, the lower surfaces on the top set of rolls being arranged to come slightly below the top of the bottom rolls, and the plate is therefore passed through the series in a serpentine manner. By adjusting the top rolls, the plate can be made to come out almost perfectly straight, but the buckle is not absolutely removed, although the work is done very effectually, especially on ½-inch and ¼-inch plates, saving many hours of hard labour. No matter how bad the plate was, it requires very little manipulation by the plater after it leaves the machine. It removes the waviness, takes out the lumps and hollows, and appears to concentrate or make the buckle more defined, and this generally towards the end, giving the plater a very good idea where to place his flatter; whereas without the machine, the plater might be hammering for hours, even before he got to the buckle. The machine appears to act more effectually when the buckle is in the middle of the plate, elongating the edges in a greater proportion than it would the middle supposing the edges were buckled.

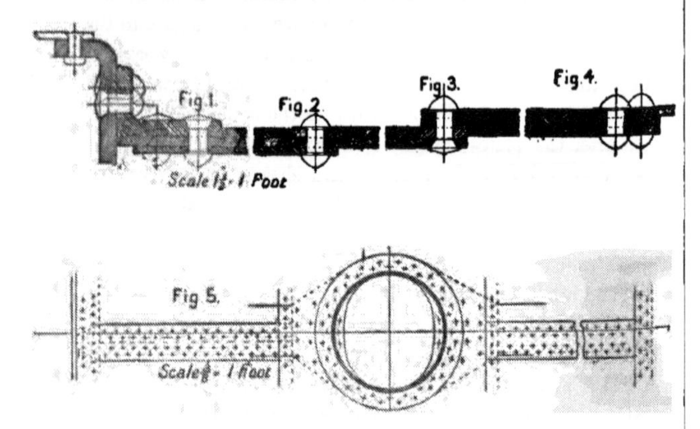

The barrel may be made telescopic, as shown in the general drawing, and of three plates, or, as in some cases, with only two, each about 5 feet 6 inches long, butting together, covered by and riveted to a weldless ring, the longitudinal seam being welded. The transverse joints are single riveted, Figs. 2–5; longitudinal seams are butt jointed, with inside and outside butt strips, and these break joint on each side of the centre line of the boiler at the top, as shown in Fig. 5. The seam joining the barrel to the fire-box casing is zigzag riveted, Figs. 5 and 6. A strengthening ring or plate ½ inch thick is riveted to the inside of the middle barrel, as per detail drawing Fig. 5, about the dome hole, which hole does not exceed 19 inches by 22 inches. The barrel plates are then taken and examined for surface and buckle, and invariably found to be sufficiently straight for machine purposes. The plates are ordered to be sheared with an extra half inch in width and length, in order to get the plate square with the template. A batch of five is then taken and the top one marked for drilling the holes, every hole required being marked and drilled at a radial or specially designed drilling machine. It is a very convenient practice to drill the first or future template half through with a ¼-inch drill, and then follow on through the batch with the full-size twist drill. If the full-size drill is used first, the point, being so broad and not actually a cutting edge, is very liable to run, and then of course, it has to be drawn with hammer and chisel. The first batch having been drilled,

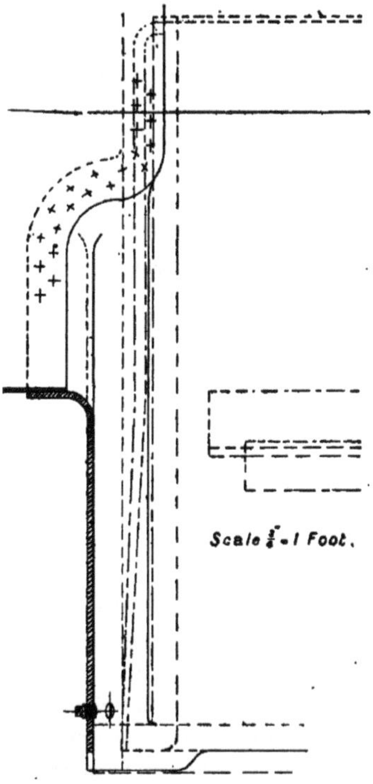

Scale ¾" = 1 Foot.

FIG. 6.

the top plate is used for the template of the next four or five sets of five, and the second plate of the first batch reserved for a future template, it being as true as the original. This mode has now been abandoned in favour of a specially designed multiple drilling machine, by which the pitch is secured by a lock upon the side of the machine, and is therefore absolutely correct. This refers to the barrel and outside fire-box shell plates, care being taken when marking out the pitch of the rivets in the transverse joints, to get the plates with increased diameters divided equally into the same number of rivet holes. Where punching is practised in using iron plates, in all cases the plates must be punched from the opposite side, that is, the one from the inside, the other —being lapped by the first—from the outside, otherwise the least diameter would not be brought together, and consequently the rivet would not completely fill the hole. It is also the practice in some large works to punch the holes of a less diameter than required, and after the plates are bent and fitted together, to rimer them out by specially designed machinery. In other works the drilling operation takes place at a specially designed self-acting shell-drilling machine, after the plates have been bent and fitted together, twist drills being used, the machine being capable of drilling both the longitudinal and transverse joints. This machine consists of a table, upon which the shell rests, like a wheel lathe chuck with dogs, which receives a motion upon the same principle as that of the wheel-cutting machine. The drill is attached to a slide capable of vertical movement, which is fixed to an upright having horizontal traverse, this latter movement only taking place as the diameter of the shell increases or decreases, the feed of the drill being accomplished by means of screw

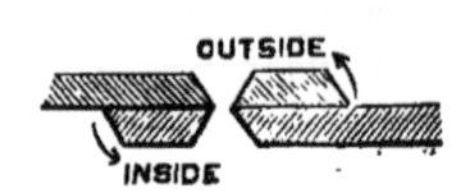

and suitable gear. The drills of this machine are generally multiple.

After the plates are drilled they are marked for lap by using the butt strip template at each end, thus:—The template is laid on and steadying pins put through the holes, the ends marked with a scriber, and then either a long straight-edge and scriber, or a chalk line is used, the plates having been drilled as far as possible to admit of shearing on two edges only, one side and one end. At the same time, the second barrel plate is marked for punching out the dome hole, which is not removed until after the plate has been bent, being held from four opposite points; and the third barrel-plate is marked for punching the clearance for the saddle or throat-plate, which is removed entirely before bending. The plates are then sheared and punched, leaving $\frac{1}{8}$ inch or $\frac{3}{16}$ inch for the planing machine. In all boiler shops there are plate edge planing machines to plane edges from 10 to 20 feet, some with turnover tool boxes, some with quick return; the former should be used wherever possible, because of the great saving of time, not to mention wearing the back edge of the tool, for there are very few side planers fitted with spring boxes to lift the tool completely away from the work upon the quick return. Others are provided with the vertical lathe arrangement, and finally, the plate edge and butt planing machine. When planing the $\frac{1}{8}$ inch or $\frac{3}{16}$ inch from the edges of the plates, the tool is ground to an angle of about one in eight, and the edge of the plate must have the inclination of this angle, according to its eventual position, to suit the fullering; that is, the inside lap is planed to the opposite inclination of the outside lap, clearly shown in Figs. 1–4 and 11–13.

The shell plates are next taken to the rolls and bent to diameter, the operation simply consisting of putting the plates through a set of three cast-iron rolls from 12 inches to

18 inches diameter, and rotating backwards and forwards, screwing the top or side roll, as the case may be, up or down, until the plates acquire the desired radius. This method, however, leaves about 4 or 5 inches at each end straight, which has to be set afterwards, generally either by making a casting, or setting a stiff plate to the required radius; then heating up the shell, dropping it into the "former," and striking a few blows with a medium heavy hammer. The method of getting the shell out from this machine, Fig. 7, is rather cumbersome: the top roll is raised up clear of the shell and wedged at one end, while at the opposite end the housing is removed and the plate drawn out. Another and a very convenient plate-bending machine, Figs. 8, 9, has the centre of the top roll in the same plane as the bottom one, the third roll being placed at the side, which is the adjustable one. This machine bends the plate to the very edge at one end, and by a very simple arrangement, shown in Fig. 8, the plate can be taken out, turned round, and the other end bent, so that the shell does not require any setting in blocks; this being the method used in bending the plates for the boiler in question. The bearing at A, Fig. 8, can be withdrawn as indicated, and then the roll swivels round upon a kind of ball-and-socket bearing at B, counterpoised by the roller and its path C. Care must be taken to enter the plates square, which is easily done by running the eye along the edge of the plate, or a line of rivet holes, and the top roll. This is especially necessary in the case of dome plates, as owing to their short circumference and stiffness they will not spring so much as a barrel plate. The plates are rolled almost exactly to radius, but sometimes, owing to little irregularities, such as hard places, buckle, &c., they need a little setting afterwards, which is easily accomplished

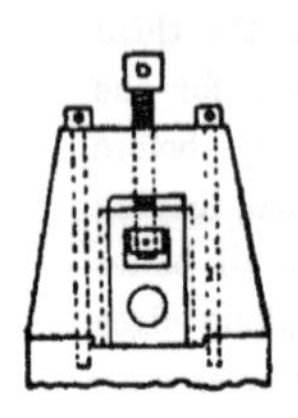

Fig. 7.

by two or three blows with the hammer. At the same time the inside and outside butt strips are dealt with, being drilled through a template in the same manner as the barrel plates, in batches of five. They are countersunk about $\frac{1}{8}$ inch deep, and all burrs removed, then marked for shearing, being sheared from the side next the barrel plate, so that the bevel edge comes next the shell for fullering. They are not planed, because the shear blades are kept in good order, and being small plates, the workman has perfect command over them, consequently the shearing is a good finish; they are then bent to the radius of the shell. The dome strengthening plates are drilled through a template, marked for shearing and punching, and then bent; these also in batches of five, and with the same template system as the barrel plates.

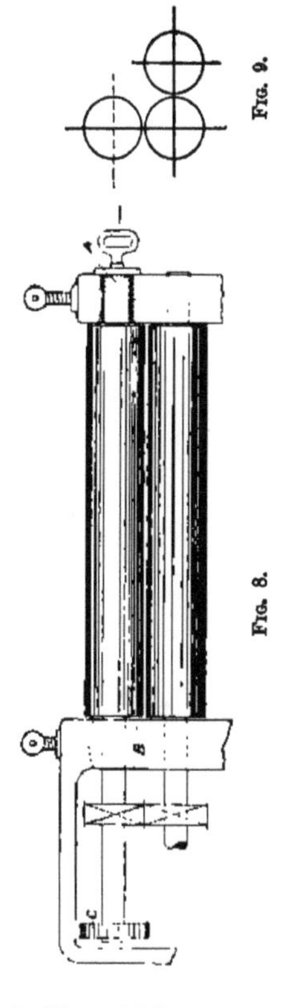
Fig. 9.

Fig. 8.

The continuous weldless ring of mild steel, from 28 to 30 tons per square inch tenacity, and 20 per cent. elongation on 8 inches, Fig. 1, for securing the smoke-box tube plate to the first shell plate, is faced, bored and turned to section, slightly heated, and shrunk on to the first barrel plate. This ring is drilled through the barrel plate from the inside, which acts as a jacket, by a specially designed drilling machine, using the opposite side to resist the thrust of the drill; or at a specially designed flange drilling machine. After the ring has been riveted to the first barrel shell, the three shells, straps and dome strengthening plate are all bolted together and riveted at one operation, Fig. 10, and it has been found that the riveting up of these two joints elongates the barrel $\frac{3}{32}$ inch, and allowance is made accordingly.

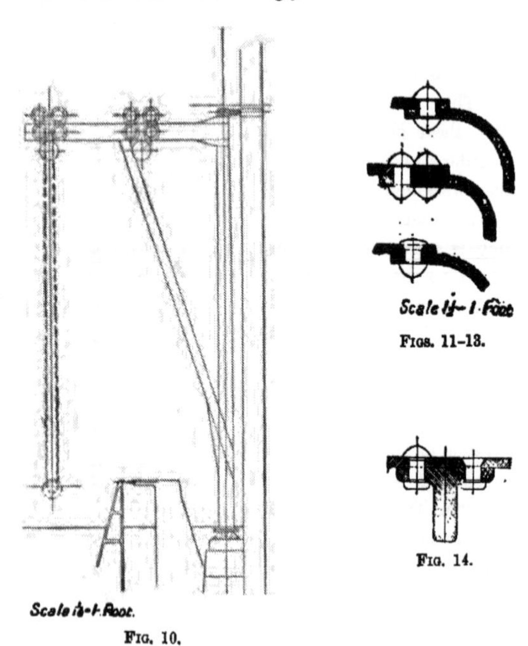

Figs. 11–13.

Fig. 14.

Fig. 10.

The sides and top of the fire-box shell are made in one plate. The front, throat or saddle plate of the fire-box shell is flanged forward and double-riveted to the boiler, Figs. 6 and 11. The back plate is flanged to 6-inch radius outside, and single-riveted to the sides and top, Figs. 12 and 13; the upper part is stayed by a T girder. A steel casting, Fig. 14, is double-riveted to the inside of the plate. A similar stay is fixed to the inside of the smoke-box tube plate; also longitudinal stays, palm stays to the copper box, and other stays where shown on the drawings. The T iron for carrying the

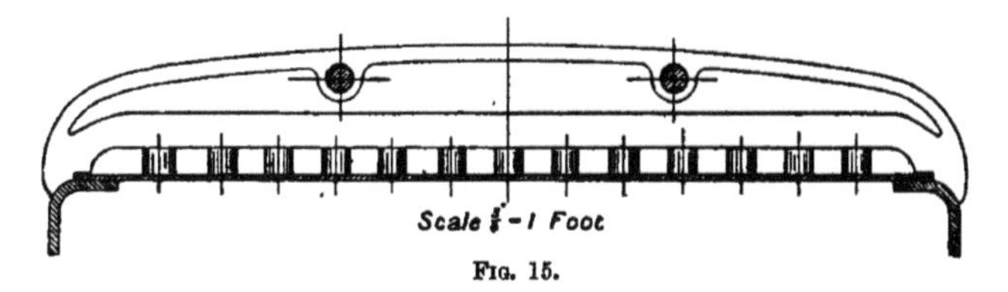

FIG. 15.

suspension links is also a steel casting, as well as the roofing stay bars, Figs. 15, 16 and 17. In dealing with the wrapper or outside fire-box shells, all the buckle is taken out, and then they are dealt with in batches of five, the same as the shell

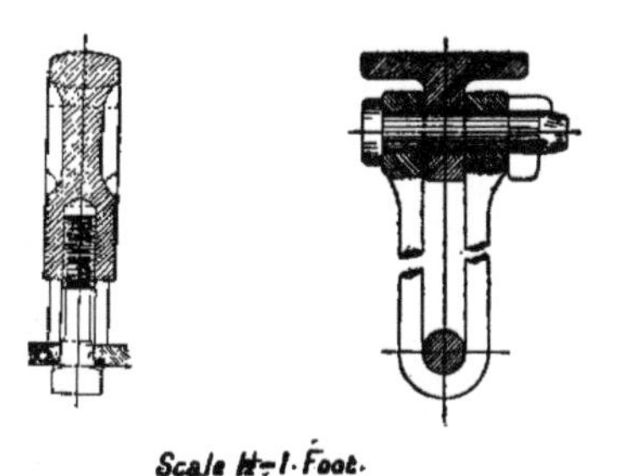

FIGS. 16, 17.

plates, every hole being marked on the template for drilling for the stays, expansion bracket, handrail, clothing studs, safety valve, roof bars, whistle stand, mud plugs, &c. After drilling, which has been done at a multiple drilling machine, with a lock upon the side of the machine to preserve the pitch accurately, they are marked, punched, sheared and planed in the same order as the barrel plates, and then taken to the plate bending rolls. The short sets to a radius of 6 feet are worked first, the plater working to the lines as shown on the sketch below.

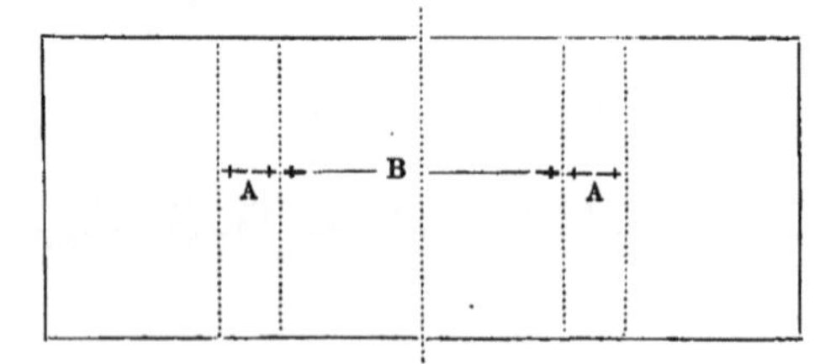

A, bent to 6 feet radius. B, bent to 2 feet 3 inches radius (outside).

The plate is then turned over and the crown radius bent, trying from time to time with the half gauge. When the machine, Fig. 8, has been used a number of times, this half

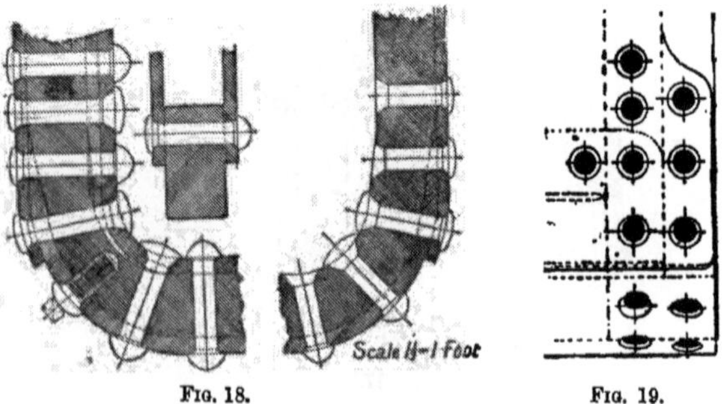

Fig. 18. Fig. 19.

gauge is not required very much, because of working the adjustable roll to previously worked-out marks. After bending, a stay of standard width is bolted between the sides to keep them from sagging the radius, and then they are taken to the drilling machine for the rivet holes between the frames to be countersunk. The foundation ring is a steel casting, shown in detail in Figs. 18, 19 and 102, 102A. After annealing it is sent to the furnace, in order to have contraction defects removed, such as squaring-up and straightening. It may be that the ends are not at right angles with the sides; if so, all four corners are heated, and a stretcher is fixed across the opposite corners diagonally, drifting it out and "paening" the corners not drifted, until all four sides become square with each other. Afterwards it is sent to the grinders, then the rivet holes are drilled through a jacket or jig at a multiple drilling machine, and the other holes, such as the corner rivet holes and ash-pan studs, at a radial drill. It is then sent to the smith's fire to get the flange plates thinned and set to the corners. It is contended by some well-known boiler-makers that locally heating a steel plate, thinning-down and hammering the corner to a bevel edge is a bad practice; consequently they use a specially designed plate-corner thinning machine, which thins off the corners very uniformly; but in this case the plate has to be punched thus—

otherwise the maker would have to be simply content with a single rivet, and therefore lose the benefit of the extra lap; whereas if drawn out they are simply sheared straight. Some firms slot a recess in the ring at the corners.

Generally speaking of flanging, it is preferable to have the flangers A, Figs. 20–29, cast in segments, frequently in halves, because they always get larger in use, and if cast thus, slack can be easily taken up by removing a ¼-inch liner and substituting a ⅛-inch. The flanging and vice blocks B C, Figs. 20–29, are cast in segments, which gives a certain amount of springiness, and if a block breaks it

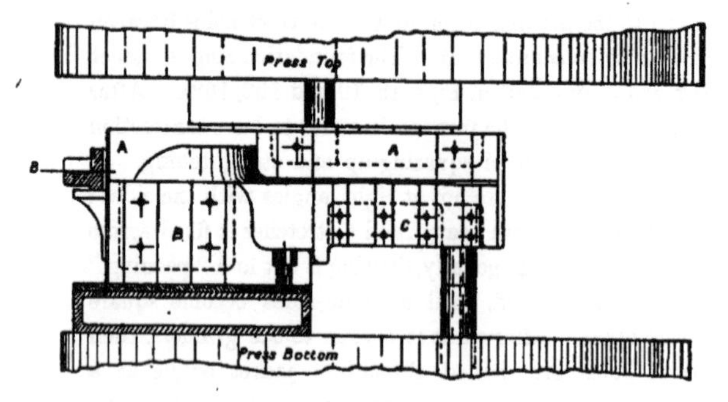

Fig. 20.

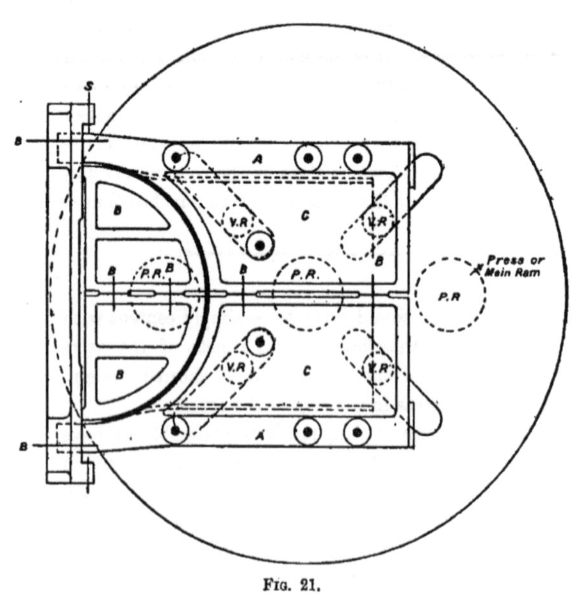

Fig. 21.

Scale $\frac{3}{8}''$ = 1 *Foot.*

can readily be replaced by casting another segment; then the curves can be easily got up by fitters where it is not

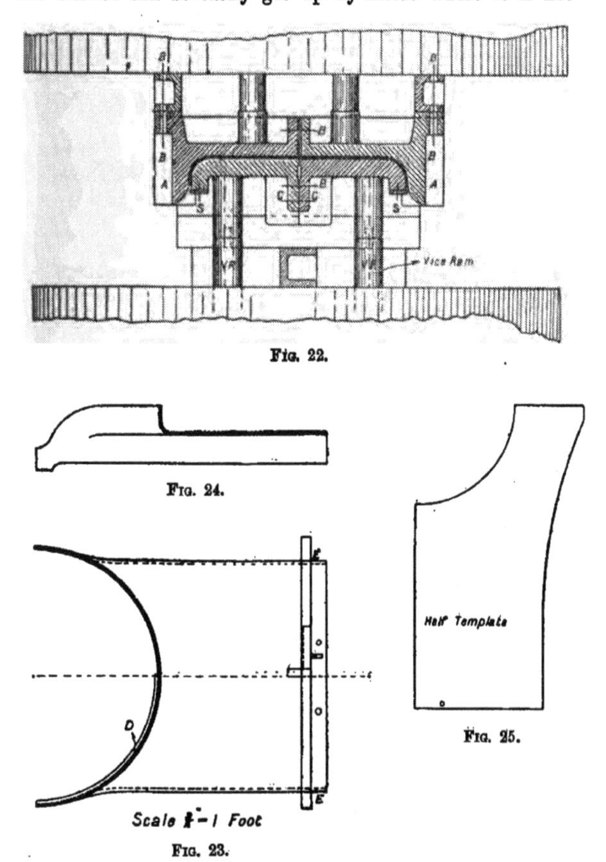

Fig. 22.

Fig. 24.

Fig. 25.

Fig. 23.

possible to turn them, the requisite amount planed off the strips left for this purpose, and the blocks then jointed together. In flanging the throat-plate, Figs. 20–25, the half template, Fig. 25, is first made, the centre line is marked down the plate to be flanged, the half template put

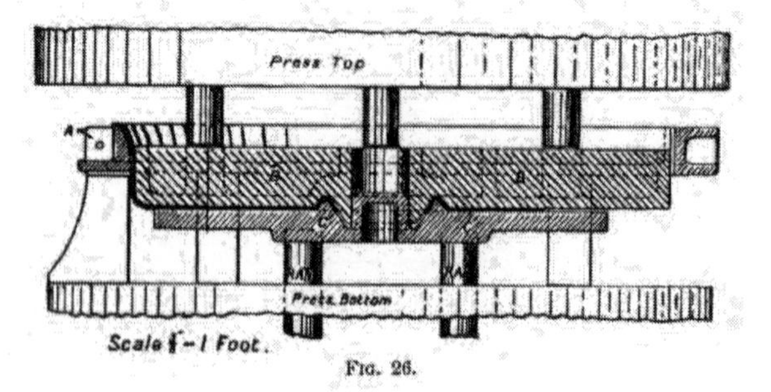

FIG. 26.

on, marked, punched and sheared, leaving a little for trimming up after flanging, because it is possible that the plate may draw one way or another, and then sent to the flanging press. The press has three main press rams acting upon the

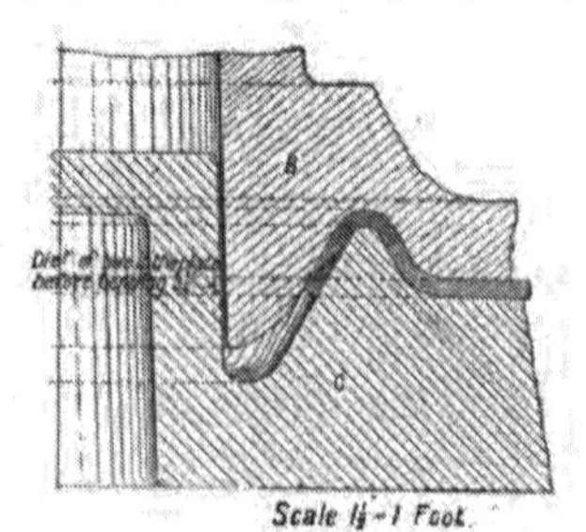

FIG. 27.

press bottom, and four vice rams, which act independently of the main press rams through apertures in the press bottom. The press top is supported and the thrust received by four bolts, and is capable of being adjusted to the required height. The flanger A, Figs. 20, 21 and 22, is securely bolted to the

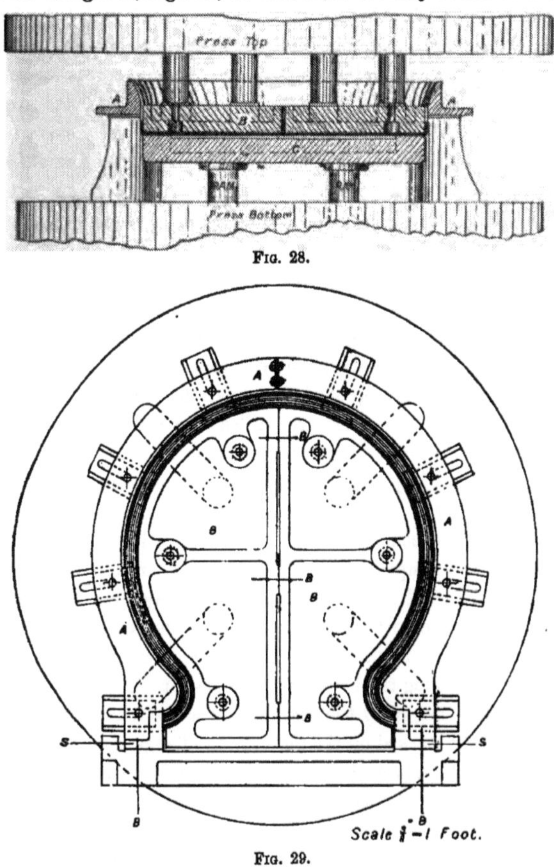

FIG. 28.

FIG. 29.

press top, using suitable packing, the throat flanging block B to the press bottom, and the vice block C rests upon the vice rams. The plate is then got to a good red heat, placed upon the vice block, centre line for centre line; also the plate has two holes punched in it, Fig. 23, to correspond with holes in the block, and steady pins are put in these to hold the plate in position as accurately as possible. The pressure is turned on the vice rams, by which operation the flange to the firebox is formed. Whilst gripped in this position, the press rams come into operation, and the throat flanger B then forms the flange to the barrel plate. In flanging the back plate, Figs. 26 and 27, motion is given to the vice rams, which forms the fire hole, and whilst holding the plate in this position, the main press rams are put into motion, forming the flange to the outside casing. The flanging of the smoke-box tube plate is similar in every respect to that of the back plate shown in Figs. 28 and 29, excepting the vice rams, which only hold the plate against the top block B whilst the flanger A does its work. The tube holes are drilled through a jacket at a multiple drilling machine, and the surface for the steam pipe flange is faced up. The finished plates and templates of the back plate and tube plate are shown in Figs. 30–35. After flanging, and contraction is over, the plate is set exactly to gauges and templates, all radii being made exactly right, the flanges squared up and the plates generally levelled. They are then marked for lap, punched and sheared, the corners thinned where required, and all holes marked for drilling.

On marking the holes off in the throat plate, first try for levelness, then carry the vertical centre line round the flange, previously marked on the plate by the flanger, and cramp on the half template at D, Fig. 23, which has been made the exact depth of the flange, making sure that the planed edge of the template is in the same plane as the inside of the throat plate. Mark the holes, and whilst this template is on, with a pair of reversers mark the first vertical hole for the joint with the outside or wrapper plate. From this hole fix the template for the vertical holes, and mark off the length upon each side; then place a straight-edge across E, Fig. 23, at these distances, and lay a square upon the centre line, the edge of the blade in the same plane as the edge of the straight-edge, and try over with a pocket square, in order to ascertain that the holes and length are square with the centre line. By this means the rivet holes in the circumference, and the

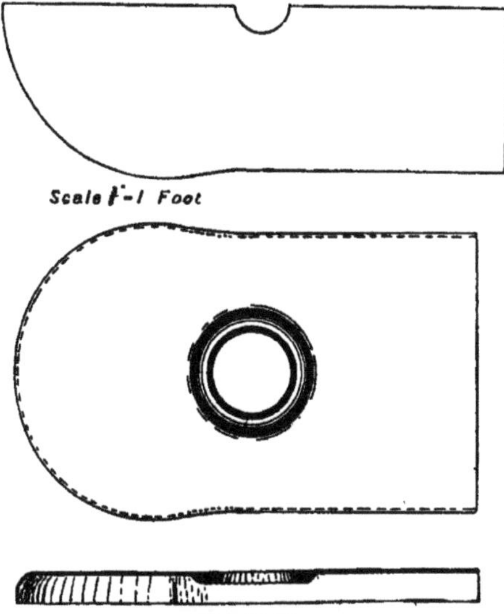

FIGS. 30–32.

vertical ones for the wrapper plate, are bound to be square with the boiler. The centre lines of the rivet holes are then marked with a surface gauge, using either the inside of the throat plate or placing it upon parallel blocks, and marking from the surface plate.

When marking the holes in the back plate, it is first, tried over with a straight-edge, placed upon a surface plate, trammelled for the centre line, and this line carried round the flange. A whole template is then cramped on, and all holes marked for stays and mountings. The plates are then turned over, placed upon trestles, and a half-gauge cramped on for marking the rivet holes. This half-gauge—it has the same

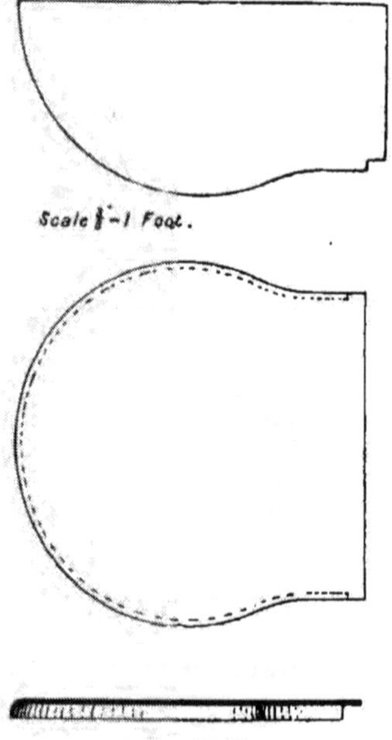

FIGS. 33–35.

section as the wrapper plate—was drilled when straight, from the same template, and afterwards bent to radius, the same as the wrapper plate; so that the holes in the back plate must come in exactly with the wrapper plate. The half-gauge is made to the right length, and this is marked upon the plate for shearing. All half-gauges or templates are made in the same manner, so that with a system of templates thoroughly carried out, it is impossible to go far wrong, and the saving of time and wages is very great. After annealing the flanged plates, the wrapper plate, the back and throat plates are bolted together and riveted, and the box is then ready for jointing to the barrel.

In jointing the casing to the barrel, Figs. 36, 37, it is first set square by means of the plumb bob A, utilising the original centre line marked for the flanger. The barrel is then bolted to the casing, being set by the aid of a spirit level, suitable packing being used to make up for the telescopic effect of the joints. The exact overall length is then determined, and B B′ is cramped on

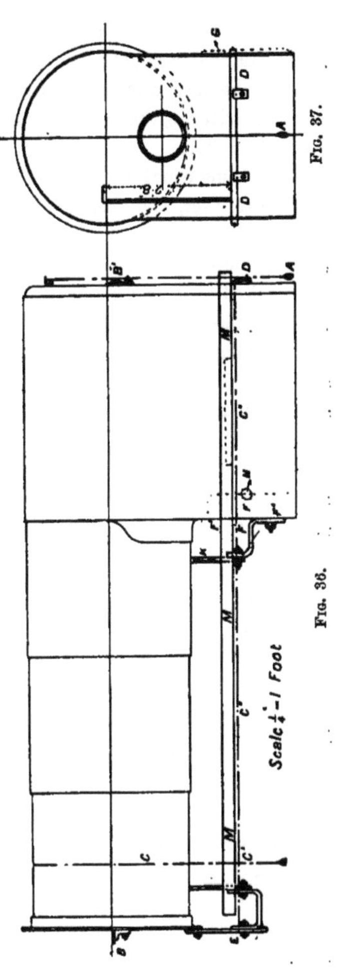

Fig. 37.

Fig. 36.

to the weldless ring and the back plate; straight-edges are then placed across, and measurements are taken with a longitudinal straight-edge. This length is made minus $\frac{1}{8}$ inch, because after riveting up the throat joint, and after the trial fire, the expansion will not completely recover itself by $\frac{1}{8}$ inch. The outside diameter of the barrel at C, Fig. 36, being equal to the width between the frames, a line is thrown over, and then the strip D, equal to the width between the frames, is fixed to the back plate, giving a $\frac{1}{2}$-inch clearance upon each side of the fire-box. The strips E are then bolted to the weldless ring, and a line carried round the boiler C″, just touching the line at C′, and equally divided at D. After trying if the box is parallel on each side with the lines, and should it be found, say, $\frac{1}{8}$ inch wide at F and $\frac{1}{16}$ inch at G, that is, a small twist upon the box, a stretcher is fixed in tension at H, and the radius of the flange paened at F^1, the same operation being repeated at the other corner for $\frac{1}{16}$ inch, until the sides of the box are made parallel with the lines and square with the framing. This paening is avoided as far as possible, but it is almost absolutely impossible to entirely dispense with it. If the box is narrow, the stretcher should be in compression and the root paened inside. To maintain the level of the plate, the root should be the portion always paened. It is generally more difficult to set in, than set out.

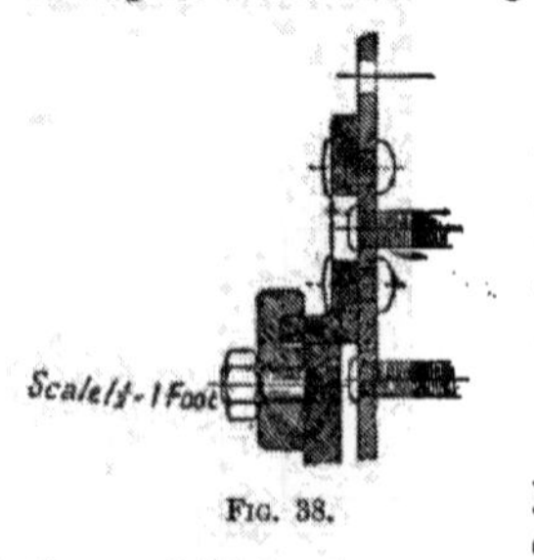

Fig. 38.

Repeated trials are made until the box is right, the length is tried over again, the joint firmly bolted, and then riveted, as in Fig. 10.

The boiler is now ready to be marked for the expansion angle iron, Figs. 36–38, and the longitudinal centre lines along the barrel. In this case the bottom edge of the expansion bracket is 2 feet 8 inches below the centre line of the boiler, and a gauge is first made, thus—

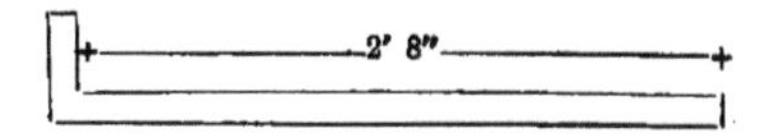

and then a straight-edge is fixed to the back plate at D, using the above gauge so that the top of the straight-edge is an exact parallel line with the cross centre line upon the back plate previously marked for the benefit of the flanger. The boiler is then levelled transversely. The radius of the barrel next the smoke-box is then ascertained, which is in the example before us 2 feet 1 inch, thus requiring a 7-inch packing between the bottom of the barrel and the plane of the expansion bracket, the radius of the third barrel-plate being 2 feet 2 inches, gives a 6-inch packing for K, Fig. 36, which packings are generally of hard wood. Straight edges are then bolted to these brackets which are capable of vertical adjustment, and then levelled, because the boiler has already been levelled transversely. Owing, perhaps, to a little twist in the boiler from riveting up the throat joint, these straight-edges may not be in the same plane. The boiler may be just a shade up or down at the smoke-box end, only a very little, which cannot be altered; consequently the middle straight-edge may be high or low, as the case may be—nothing to matter as a rule, otherwise there would have been a serious error. A line is carried from the fire-box straight-edge to the smoke-box straight-edge, having the thickness of, say, two or three pieces of paper between it and the straight-edges, which gives a better chance of seeing if the middle one is in the same plane; then adjust, when the exact elevation or depression of the smoke-box end is ascertained. Afterwards place a long straight-edge across the three and level the boiler longitudinally, and mark along the casing for the expansion bracket, the holes for which have been drilled at the same time as the stay holes. Supposing the line for the edge of the bracket is, say, $\frac{1}{16}$ inch out from the required position, it is passed; but if more, the boiler must be canted over, whichever way required, until the error is divided, this requiring to be done in something like 2 per cent. of the total boilers made. The brackets, Fig. 38, are steel castings, with the holes for clearing the stay heads cast in. They are bolted to the casing through these holes, and then the foreman tries them over, marking them and the casing with trammels for the riveter to work to, who notes that they do not shift in riveting. Four holes are marked from the inside of the casing, and the rest are drilled through a jacket. Having marked the expansion bracket, the longitudinal centre line along the barrel has to be marked, using the above 2 feet 8 inch gauge, only the reverse way. Lines are thrown over the barrel, and the straight-edge M brought up to them, the gauge otherwise would cant over, and consequently not show the exact length. Each end of each section is then marked by scribing in between with shorter and more convenient straight-edges, at the same time carrying this line to the weldless ring. The smoke-box tube-plate is then cramped on, the centre line of the plate to the centre line of the weldless ring. A straight-edge is then held upon the bottom edge of the plate, which has been previously planed. It is then levelled, and the rivet holes marked with a ring punch.

The dome is formed from an oblong plate of best iron or mild steel, welded at the seams and flanged at the bottom, with an angle iron ring at the top, and fitted with a wrought-iron or mild steel cover. The flanges of the dome and cover are faced, so that steam-tight joints are made. It is marked from template and sheared, allowing from $\frac{1}{2}$ inch to $\frac{3}{4}$ inch for welding the contour to form the flange for riveting to the boiler, and four holes being punched. The inside edge is planed bevel for the scarf, the plate is then bent to radius, and rivets put in to hold it together whilst welding. After fixing the welding block and stand, with its platform for the smith and strikers, the porter bar is clamped to the dome shell, Fig. 39. If there is too much lap for the scarf, the ring after warming is paened on the outside, and if too little, on the inside, until the required lap is obtained. The first is a nice wash heat, the weld being shut about 6 or 8 inches from one end, a fuller being used to bring the outside lap upon the bevel edge of the inside, the scarf being off. The welds are generally "shut" at two heats, applied alternately to a forward portion and a hind one, but if not properly welded up a third is required, the straightness and radius of the dome being watched throughout. Great care is taken with the heats, this being necessary,

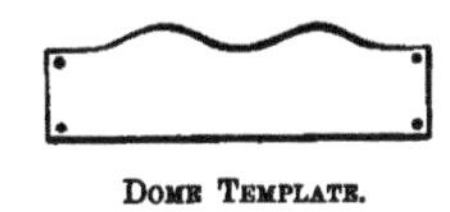

DOME TEMPLATE.

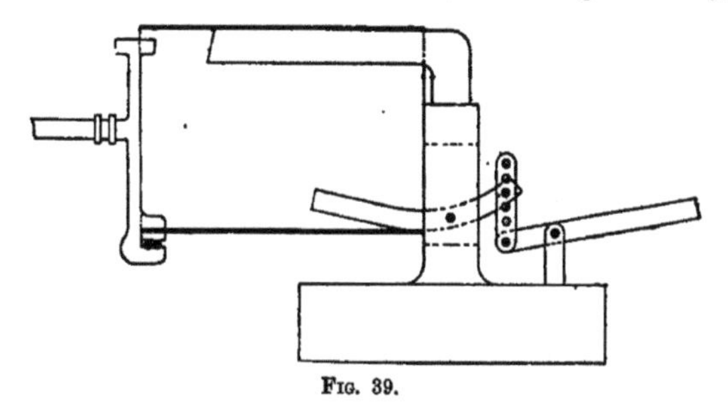

FIG. 39.

because the outside lap heats up quicker than the inside; although a fire-brick is placed over the portion to be shut, which retains the heat well. After the seam has been shut along its whole length, the end to be flanged has a piece of bar iron welded across the joint; after this, it is never known to open during flanging.

The flange is formed in a block, Figs. 40, 41, 42, about one-third to one-quarter at a time, after which it is set per-

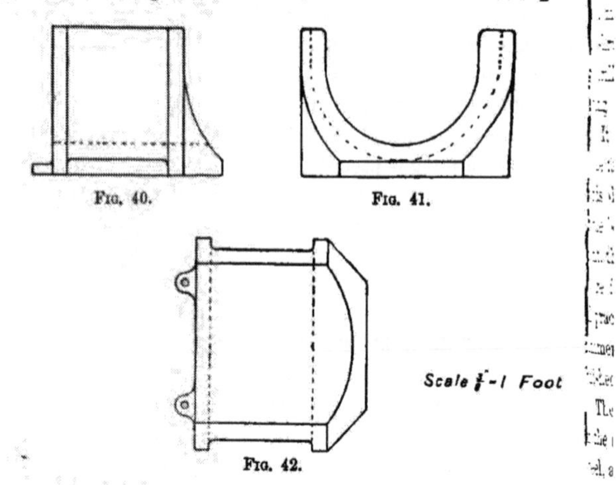

FIG. 40. FIG. 41. FIG. 42.

fectly right in relation to its axis and sides, using the gauges upon a block similar to Fig. 43. The dome is then placed upon the template, Fig. 43, and squared up from the surface plate; four holes being set out and the rest drilled through a jacket. For this purpose the dome is fixed upon a frame on an axis, in order to get the holes at right angles with the flange.

DOME GAUGES.

The seating for the cover may be flanged, or formed from an angle iron, which is bent to radius on a block at the furnace, the ends being bevelled to form a scarf; and in welding up, a piece of square iron is let in at a welding heat, bent over upon the flat, and completely shut, the same precautions being taken as to heating. All cases are finished accurately. It is then skimmed up for entering the dome, which is warmed and shrunk on. The two are drilled together whilst

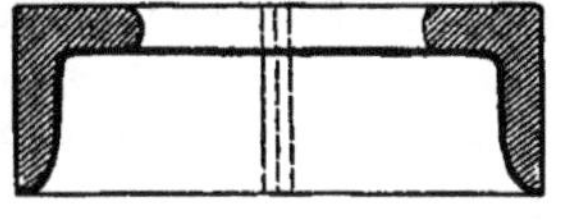

resting upon a roller frame, and afterwards riveted up. It is then cramped to a block in the lathe, squared to the face plate, and the face of the angle turned up; afterwards drilled, tapped with a lightning tapper, and riveted to the boiler. It is found by following out the above treatment, that although the dome has never been near the barrel before, it is almost absolutely correct, certainly sufficient for all practical purposes. The covers are dished at the steam hammer, the tup and anvil being convex and concave, and finished by turning the faces and edges.

The safety valve seating is made similar to the angle ring in the dome, using the required section of best iron or mild steel, and flanged to the radius of the outside fire-box shell; now, however, speaking generally, a steel casting is used. The holes are marked from a template, similar to the dome template, Fig. 43, the scriber D being used to set out

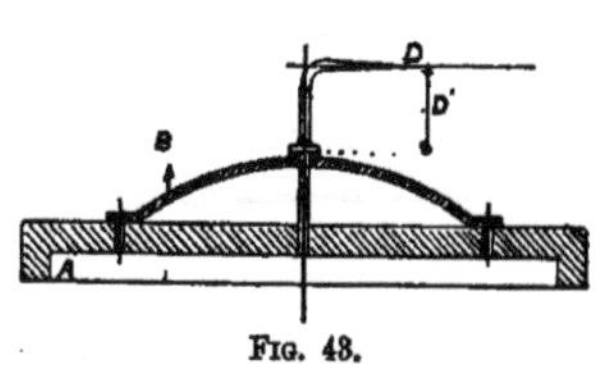

Fig. 43.

how much must be turned from the seating. Four holes are marked, and the rest drilled through the template or jacket.

The inside fire-box shell is made from the best quality copper plates, having a tenacity of from 13 to 14 tons per square inch, 45 per cent. elongation on 8 inches, and 50–60 per cent. contraction of area; the stays of about 14 tons per square inch, 45 per cent. elongation on 8 inches, 55–60 per cent. contraction of area, also bearing the test of being bent double, cold, without any signs of cracking. The crown and sides are in one plate similar to the outside casing, and the tube plate is widened out, forming a pocket or "joggle" on the side plates to allow a wide spacing of tubes. The roof is stayed with eight girder stays of good well-annealed steel castings, Figs. 15–17. The copper stays are screwed tightly into the fire-box and shell plates by hand, in order to be absolutely certain as to the tightness of the same. In cutting off the stay ends, care is taken not to injure the threads, and a projection of about $\frac{1}{2}$ inch is allowed on the inside of the fire-box, and $\frac{3}{8}$ inch on the outside shell, after which the stays are carefully laid over to a neat job, inside and out, without being snapped, with hammers not exceeding 4 lbs. in weight. Snapping is a bad practice, as it indents the plate, forming receptacles for verdigris and corrosion, which eventually entails the use of stays as much as $1\frac{1}{4}$ inch in diameter. Basic steel stays are also used, with reputed success.

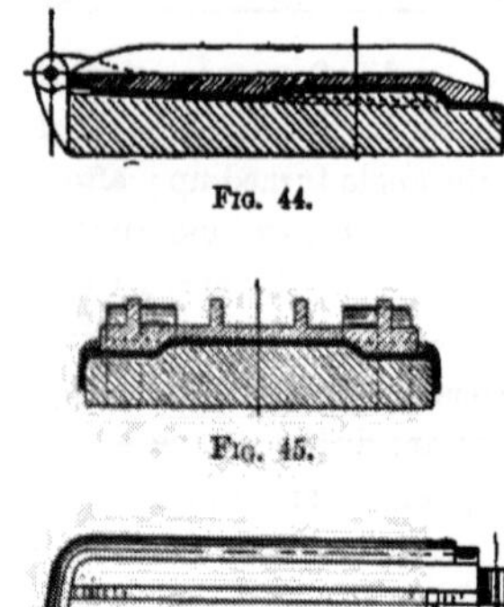

Fig. 44.

Fig. 45.

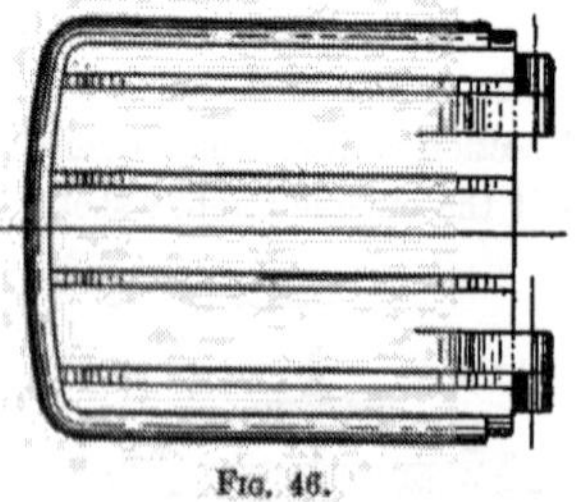

Fig. 46.

The copper plates are treated in the same manner as the steel plates as regards drilling, planing, &c. They are flanged by hand, as represented by Figs. 44–46, and the following is substantially the universal method of dealing with them. Five heats are required to complete the back or fire-hole plate, viz. two for the flange, two for the fire-hole and one for annealing by quenching. A third heat is very seldom required for either process, save upon rare occasions, and for general practice it is not necessary, but owing perhaps to some delay, or the flange being not quite down upon the block, it is sometimes rendered imperative. The tube plate requires four heats; at the first, the corners are worked to avoid concentrating the metal, which would otherwise pucker

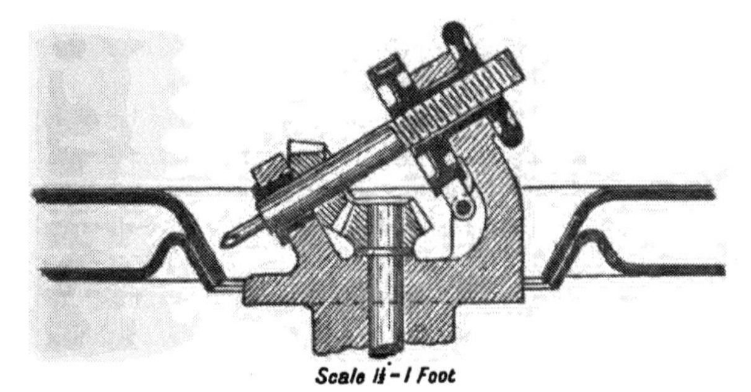

Fig. 47.

up, and as much as possible is done to the top flange. After the second heat further progress is made, usually the flanges being finished as near as possible with mallets. The third heat is devoted to finishing with flatters which are about 4 or 5 inches square, and the fourth to annealing by quenching. After flanging the copper back plate it is bolted to the outside back plate, and the fire-hole rivet holes drilled. A machine has been designed by Mr. Webb for this purpose, Fig. 47. After the inside fire-box crown plate has been bent to radius, the joggle is set in by heating and hammering into a block having the same joggle as required by the plate.

The set commences as far as possible from the edge, and near the second row of stays, ensuring that the plate is not drawn, and that the rivet holes will come almost exactly central with the centre line on the flange of the tube plate. These holes are opened out by a hand drill. The advantages of the joggle or pocket are very doubtful, and in many cases it is being dispensed with. The three plates are then bolted together, the corner rivet holes opened out, the foundation ring put on, and the corners bedded to it. The roof stay bolt holes are opened out with a special tool having the required taper, as at *a* in Fig. 16, which clears out the hole until a facing cutter just skims up the inside of the crown, to give a good bed for the roofing stay bolts.

The roof bars are then fitted on, the clearance between the top of the box and the bottom of the stay for the ferrule space having been previously planed away. The ends are then marked for machining, to bed on the corners of the box and to bring the ferrule distance right, by using an ordinary forked scriber. The bars are squared up to centre lines, the holes marked for the shackle cotters to template, the holes drilled for the stay bolts through a jacket at a multiple drilling machine, tapped with a lightning tapper, and bolted up with a 6-foot key.

The inside box is then lowered into the outer shell. In order to get it far enough back to enter the fire-hole door flange, the foundation ring is removed. The palm stays and other odd rivets having been put in, it is ready for being tapped by the quadruple tapping machine. This machine, which is shown in Figs. 48–50, consists of a framing, secured by angle irons to the foundation ring, and is entirely rope-driven, each tapping stock being driven separately. The main driving rope from the counter-shaft is 1 inch in diameter, and passes round the pulley A, the power being transmitted by 1½ inch by ½ inch belts to the corner pulleys

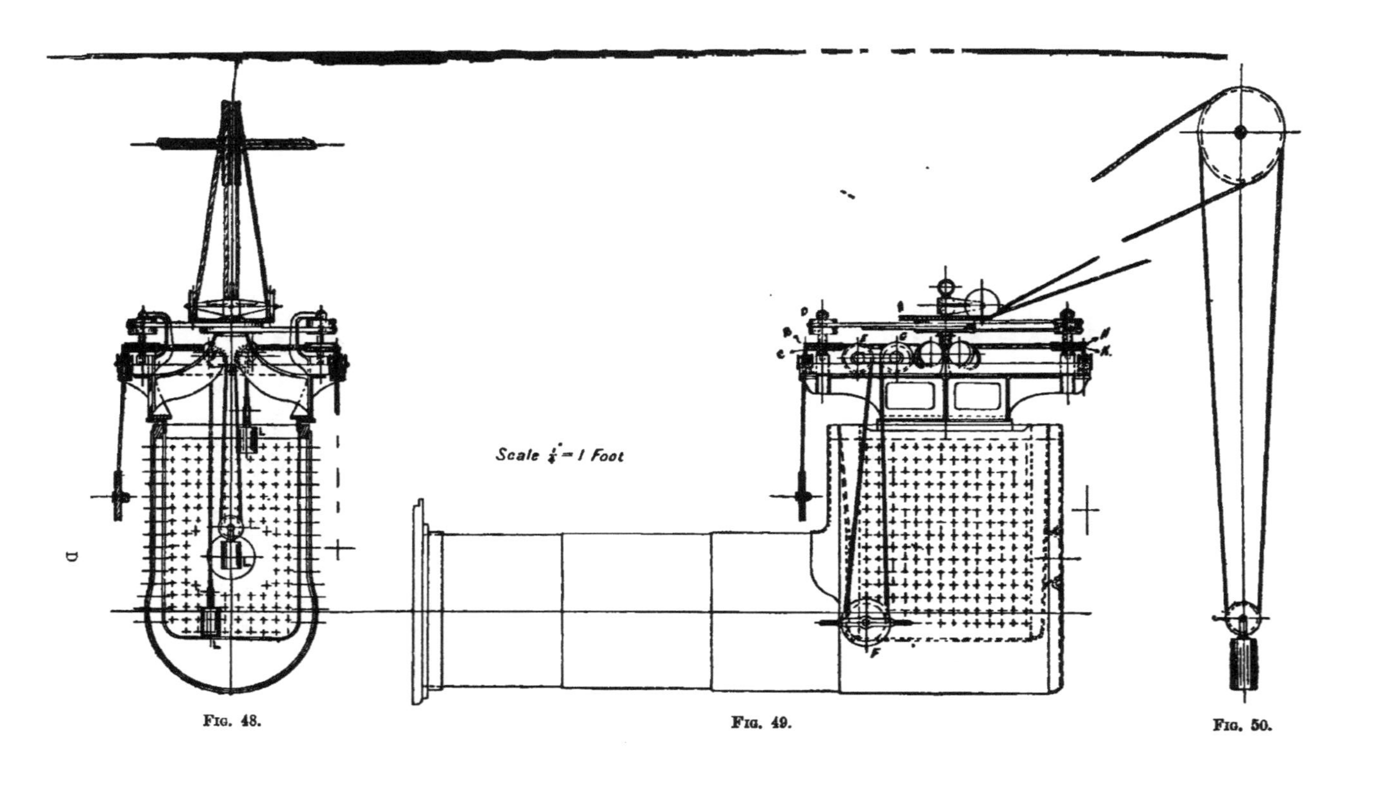

FIG. 48. FIG. 49. FIG. 50.

and this in its turn to the tapping stocks by a ¾-inch rope. As the course of the rope for each stock is similar, a description of one will suffice. B is a fast pulley driven by D, C being a loose pulley. The rope leaves B, passes round E to the tapping stock F, thence to the loose pulleys G and H, K being fast, from H to the centre of the machine, over a loose pulley, and round one of the counter-weights L to B again. This valuable labour-saving apparatus was designed by Mr. Aspinall. The roof-bar shackles are put in and cottered up under slight tension.

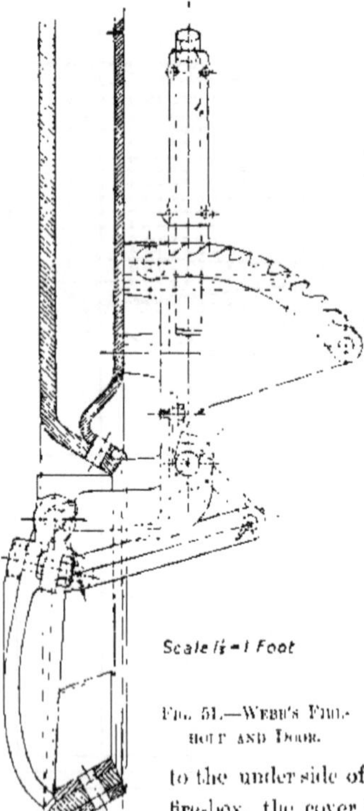

Fig. 51.—Webb's Fire-hole and Door.

The wash-out door is a well annealed steel casting, riveted to the under side of the boiler in front of the fire-box, the cover being made with a cone joint. Brass taper mud plugs are placed at the front and back of the fire-box, and in the smoke-box tube plate. The boiler is now ready for the mountings, the caulkers having done their work at the proper period, following upon the riveters.

The copper piping for the ejector and injectors, &c., is

first put in, drifted and ferruled; then a specially designed surfacing machine is cramped to the fire-hole, and the surfaces got up for the regulator gland, injectors, ejector, jet, water gauges, &c., and these bedded to the machined surface. The fire-hole door, Fig. 51 (Webb's), and hand-rail studs are fitted into their places, working to the centre lines of the boiler. The regulator steam pipe elbow is levelled in the dome, and the copper pipe with the cone on it is bolted up to it, then a coppersmith fixes the flange for the tube plate joint ready for brazing. The tubes, if copper, must be solid drawn and seamless, perfectly sound and well finished, free from surface defects, and also capable of withstanding expanding and bending, without showing the least sign of splitting or cold shortness whatever.

The ends should not be annealed, but left "hard" or "half hard" throughout, because, if the ends are annealed, the junction of the hard and soft metal becomes a plane of weakness, and the tube invariably collapses there.

The thickness must be 10 I.W.G. for 12 inches from the fire-box end, and then taper from 10–12 I.W.G. in a length of 18 inches, the remainder parallel, and 12 I.W.G. thick; to be swelled $\frac{1}{16}$ inch at the smoke-box end for 3 inches, to facilitate withdrawal.

Slight deviations may occur, e.g. cambered tubes may be required, which should be specified when ordering.

The weight per lineal foot is as follows:—

$1\frac{5}{8}$ in. outside diameter,	1·98 lb.	A maximum of 10 per cent. above each, and a minimum of 5 per cent. under will be allowed.
$1\frac{3}{4}$,, ,,	2·15 ,,	
$1\frac{7}{8}$,, ,,	2·31 ,,	
2 ,, ,,	2·47 ,,	
$2\frac{1}{8}$,, ,,	2·63 ,,	

They must be free from dirt, inside and out; each tube must be branded, and also capable of sustaining an internal pressure of 800 lb. per square inch and an external of 250 lb. per square inch. They are put in, expanded at each end, and ferruled in the fire-box. The longitudinal stays are bolted up, using copper washers between the nuts and the plates. The dome and safety-valve seatings have a finishing cut in their places, a small machine being designed for the purpose, this having been found necessary because the surface obtained from the lathe was irregular owing to vibration.

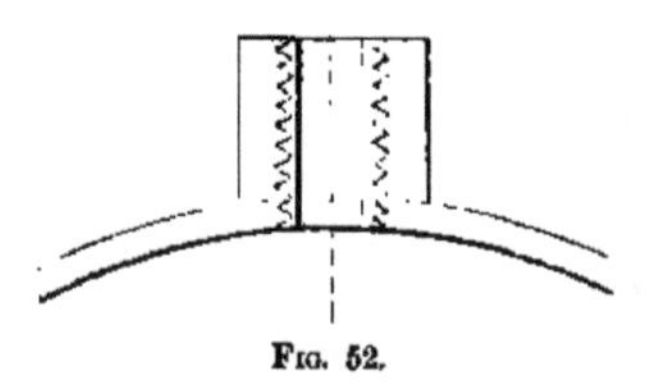

Fig. 52.

The whistle-stand joint is made by screwing a piece of piping through it and the ½-inch plate, steam-tight, as Fig. 52. The fusible plugs and brick arch studs are put in, and all joints made with a fine seam of thin red lead. Water is then put in and warmed, afterwards pumped up to 200 lb. per square inch, then drawn off until the gauge glass is left just full, and the boiler fired up for five hours, steam blowing off at 160 lb. per square inch.

SECTION II.

THE FOUNDRY.

PART I.—IRON.

DURING the past few years, the encroachment of steel castings upon the sphere of pig iron has been very extensive, and perhaps this statement may also forcibly apply itself to the forge and smithy. In order that castings may be considered first-class, it is requisite that they should be quite compact and sound. This is of greater importance in specifying castings in foundry practice, and is more valuable in determining their strength than the quality of the mixings from which they are run; therefore in all cases castings are designed with a view to free themselves from blowholes and dirt, heavy sections being cast uppermost, so that there is not any shrinkage due to lack of feeding. Fillets are well attended to, because, instead of being a factor of safety, they may become the reverse. They should be curved, because there is then less liability to crack than when made straight; and frequently light sections of a complicated casting are strengthened, in order that internal strains may be equalised, as well as to avoid draw, caused by unequal contraction. At the same time due consideration should be given to the crystallisation of the metal, the contour of all castings being as far as possible free from abrupt changes.

The difficulties met with in the iron foundry, as compared with those in the steel foundry, are comparatively nothing. The mechanical properties and constructive uses of iron castings, set them entirely apart from steel, but generally speaking, the one element, carbon, characterises the quality of each. The greater the content of combined carbon, the harder and less fusible the pig; the greater the content of graphitic carbon, the softer and more fusible, and cases are known in which silicon up to 3½ per cent. has been added to white irons, resulting in good strong soft castings. Phosphorus of course, renders the metal more fluid, but the resultant casting brittle. Remelting iron up to a certain extent improves its tenacity and transverse strength, but beyond, these properties suffer, and its resistance to crushing and hardness increases.

The tensile test is the most trustworthy for iron castings, as its results are much more favourable to a good working foundry mixture than the resistance to crushing, because frequently the best crushing tests are quite unsuitable for foundry use. Commercial testing is generally made upon bars 3 feet by 2 inches by 1 inch without being machined, specifications requiring from 25 to 32 cwts. as a central breaking load, allowance being generally made for blown and scabby samples in such terms as, "Should any of the tests be lower than 1 per cent. of the above strains they will be rejected." Some American specifications state that the bars shall be cast horizontally, the pouring to be done at two points simultaneously, the gates being one-third the distance from either end. When cast otherwise the method must be accurately reported to the purchaser. Dry sand moulds are used, the tensile and compression test bars being taken from the transverse bars. Bottom-poured bars are objected to, because they chill readily, through the metal constantly coming in contact with a higher and colder portion of the mould, and everything points, as far as the author is aware, to top pouring as being the most beneficial, especially with a good head of metal, for then the tensile result may be improved from 5 to 8 per cent. Generally the tensile tests do not exceed 8 or 9 tons per square inch, but are always specified to exceed 7 tons per square inch, and the author has tested samples of general iron up to $11\frac{1}{2}$ and $12\frac{1}{2}$ tons per square inch; cylinder iron to $13\frac{1}{2}$ and 14 tons per square inch; each having very good fractures and being good workable metal. The tensile tests are always made with shackles having spherical seats, as the ultimate strength is very much affected by the non-axial position of the sample. It is frequently asserted that American castings are far ahead of English, and probably taking them weight for weight and sectional area, they are stronger; but at the same time, the strongest and best castings can be obtained from the English foundries, provided the purchaser is prepared to pay the price. Hence, the skill of a foundry manager is taxed to a great extent in blending certain brands of pig of which he has practical and analytical knowledge, to produce the class of casting required, embodying the requisite hardness or softness, with strength and toughness. This is carried out to such an extent in some foundries, that the selected brands are melted together, cast into pigs, broken, the best fractures reserved, and at the same time test bars are cast, in order to gain. information for this special work. Should a contract be let at a very low figure, stringent means must be taken to keep the founder up to specification, and then, with all care, an inspector is not absolutely certain that his transverse bars are of the same quality as the castings, or even if they are from the same heat, therefore such samples should form part of the casting, and the inspector should see them broken from that casting, or, as in the case of girders and similar work, from 2 to 4 per cent. extra should be cast, and then selections made from the bulk, tested to destruction, and should the selection break below the specified load, the bulk should be condemned. It is often advisable, in these deflection tests, to note the ratio of the first permanent set to the ultimate load, which is perhaps sometimes a fine accomplishment, and rewarded by curious results.

In moulding, many of the operations are common to all castings. Patterns should have good draw, and be provided with draw screws or irons; the rapping holes and the general arrangements for loosening should also be well attended to. Core prints are not absolutely necessary in all cases; where they are large and rest upon the bottom of the mould, they can be set in position by measurement, and their weight, or the combined weight of them and the top part of the mould, will keep them in position, but usually cores are fixed by means of prints, and justly so, because they will save time and the risk is infinitely less. The cylinder patterns and large pipes are lagged up, which is the correct way, jointed longitudinally and accurately dowelled. A large piece of timber, well seasoned and dry, is not always available, and there is a greater amount of warping and shrinkage than when lagged up with 1½-inch timber. In the end it is cheaper to form the shape between the two cylinders, by a shaping core, because if this was formed by the pattern, it would make the pattern very weak where it requires to be strongest, not to mention the use of gaggers and irons.

The facing sand for large castings consists of red sand, road sand, coal dust and horse manure, whereas for smaller work it is of a very much simpler composition, viz., red and black sand only, whilst for core-making a loam of red sand and manure is used; but if the red sand is not so open as usual, a quantity of road sand is mixed in, and riddled according to the nature of the work. Dry sand moulds should be well dried, otherwise they will scab, and for this reason the green sand moulds should not be rammed too hard, consistent with swelling. Gaggers, studs and chaplets are well set, and attention is specially given to venting, accurately illustrated by examples to follow. It is a question in the minds of some foremen, as to whether risers should be open or closed when casting. If they are closed, the iron falling into compressed air as it were, cannot fall so heavily, the metal cannot rise so fast or seethe against the sides of the mould, which will prevent scabbing; but at the same time there should not be any noise created by the escaping gases. This question should also be considered at the time of casting, with the heat of the metal, and whether poured from the bottom or top. In order to facilitate the removal of impurities, skimming gates or cores are sometimes used, or, as in the case of cylinders and other large castings, a large basin is used, which is always filled before the metal is allowed to flow down the leaders, which prevents all dirt and scum from entering the mould. When casting in green sand in the foundry floor, the bottoms are carefully rammed and vented, the surfaces not too hard, as they do not always give the best results, and simply tapping a pattern into loose sand should be watched and condemned. Many castings of this description have their contour spoiled by an ugly joint, and the best judgment should be exercised in making irregularly parted moulds. If a joint can be made so that the gaggers can be squarely and easily placed, and so rammed with as little sand under the gaggers as possible, that joint has the best chance of being successfully lifted.

The general arrangement of the foundry is obviously a matter of great importance. It should be spacious and lofty, so that all gases, smoke, steam, &c., which are always generated at casting time, may readily get away; light is also a factor of great moment. Generally, a certain portion of a foundry is exclusively devoted to plate moulding, and is traversed by a light overhead hand-power crane, and the floor is arranged according to the space and quantity of work required, there being straight runs, each divided by a pavement of cast-iron plates. The divisions should be of sufficient width to accommodate two rows of boxes, with a mound of sand between them, the width corresponding to the class of work to which the space is devoted, such as engine brake blocks, waggon ditto, fire bars, top and front steam chest covers, axle boxes, &c. The moulding boxes, (having the lugs at the ends instead of the sides to economise space), should be well-fitting, the corresponding halves being rammed, and brought together by their pins. The advantages of plate moulding cannot be over-estimated for quality, rapidity of work and cheapness. Wherever there is a repetition of a single article required, it should be put on to a plate pattern, either in iron or brass, and when making the wood pattern for this purpose, it must be remembered that two contractions are to be allowed for.

Plate work is carried out to such an extent that the patterns may include such articles as flanges of all descriptions, hydraulic or otherwise, tube plugs, coupling rod bushes, brackets, details of the water pick-up and fire-hole door, grids, washers of various descriptions, small grooved wheels, &c., which will be illustrated further on. Economy is derived from the fact that a special class of workmen may be reserved for this class of work; also because the boxes are not brought together until the mould is finished, and the runners are not cut by hand, because they form part of the pattern in the plate. The boxes require to be rammed fairly well, or the casting will swell. The vent wire is not absolutely necessary in all cases, and very few castings are lost from scabbing. It is used for the larger portions of work, such as steam chest covers, engine brake blocks, &c.; but with such work as fire bars, small brake blocks, chairs, &c., it is not as a rule required.

The larger descriptions of moulding, such as foot plates and a variety of general work, should be done in that portion of the shop where there is suitable lifting tackle, and provision should be made for transferring heavy ladles of metal from one portion of the shop to another.

The heaviest portions of dry, green, loam, and open sand moulding should be grouped together, and there should be convenient pits for pipe work, the whole foundry being also supplied with hydraulic jib cranes where necessary. The blower, wheel moulding, and other machines should be placed in the foundry fitting shop, and the fettling shop should immediately adjoin the foundry. All castings pass over the weighing machine to the stores. The entrances to the drying stoves, which are either heated by producer gas or fired, and near which, as far as possible, all the dry sand work is done, are flush with the outer wall in all modern foundries, the stoves themselves being outside. The tapping holes of the cupolas for heavy castings are also flush with this wall, and landers bring the metal into ladles at convenient crane distance from the wall; whereas the tapping holes for shank ladle work are sometimes placed upon the opposite side in the breast, and this is generally found to be very convenient. Arrangements are generally made that all sand, coke and pig, shall be brought up in such a convenient manner that the sand store is immediately adjoining the grinding mill or disintegrator, and the coke and pig in stacks about the cupola, being raised to the charging platform by means of hydraulic hoists, where natural advantages do not admit of this being on a level with the various stacks.

It is advantageous to have two cupolas, and work them alternately, the common sizes being from 2 feet 6 inches to 4 feet, and the total height of the blast stacks varying according to the work and nature of the fuel. The positions of the tuyeres and charging holes are important. In the former case the position is determined by the class of work to be done. Where heavy castings are made, and consequently a large body of metal is required, the tuyeres are placed high, the object being to have a large body of metal in the bosh, to retain the heat, also giving a greater facility for the different brands of pig and scrap to thoroughly blend, melting more and running longer heats. Where small castings only are required, the tuyere is placed lower, and has the advantage of saving fuel, but the same depth of fuel is required over a low tuyere as over a high one. The position of the charging hole is most important, because the higher it is in reason, the less proportionately the consumption of fuel will be; for this reason high cupolas hold the heat, make the iron hotter, and melt it faster. This can be practically exemplified where convenient, by raising the charging hole for two or three charges. It is evident that the more charges there are in the cupola, the greater amount of heat will be absorbed which otherwise goes up the blast stack, and consequently when the charge comes to the melting stage it will melt quicker and hotter.

The fuel should be good hard coke, as free as possible from sulphur. Speaking generally of coke, the heating power may as a rule be taken to approach somewhat closely to the content of carbon. The purest samples of coke will only contain about ·5 per cent. of sulphur, whilst the most impure will have about three times that amount present. It should be as free as possible from black ends, because these contain more of the volatile matter existing in coal, and they are more porous, absorbing a greater quantity of moisture when quenching, than dense coke. Good coke should not exceed 1·25 per cent. of volatile matter besides moisture, and the lower the content of volatile matter, including water, the less will be the loss of heat. Ash has a detrimental effect, because it is fused, consequently more limestone is required, and it probably brings in the sulphur; however, a small increase in the content of ash may be to a certain extent ignored, providing that the mechanical properties of the coke are such that it will withstand a good crushing load, is hard, and almost of a metallic lustre. Soft, tender, friable coke must be avoided, because it cannot withstand the weight of the charges, and is more easily affected by CO_2, consequently is consumed faster and melts less metal. To those who have charge of cupolas this result is well known; with the former they may probably finish charging several hours before the latter, and then not as great a quantity of metal melted. The analysis of coke may vary in carbon from 87·5 to 95·5 per cent.; sulphur ·3 to 1·5 per cent.; ash 2·5 to 10 per cent.; moisture 1·6 to 3·5 per cent.; and some cokes may be found dry; but as with steel, the chemical properties should always be bracketted with the physical, otherwise it is useless. The quantity consumed is about 1·5 to 1·75 cwt. per ton of iron melted, or 2·5 to 2·75 cwts. per ton of castings passing over the weighing machine, which is about equivalent to American consumption; and the quantity of iron melted in a 4-foot cupola is about 3·5 to 3·8 tons per hour, including lighting up, a fair day's work being from 35 to 40 tons per ten hours' shift. But this of course varies according to the work in hand; as much as 70 tons may be got out in a twelve hours' shift with continuous working, there being a greater economy in fuel by having heavy charges. The foundry foreman should keep a strict account of iron melted, castings passing over the weighing machine, and the average consumption of coke, thereby always having a check upon the cupola account.

In preparing the cupola, it is first well cleaned out, especially for 6 or 8 inches above the tuyere holes, using hammer and chisel when necessary, cleaning the whole of the inside carefully without removing the glossy skin; all furrows should be filled up with ganister, and under good management the lining will wear very evenly. The bottom is then made with moist black sand, which is well rammed, and afterwards sleaked over with clay wash or blacking, it being slightly inclined to each tapping hole. The tapping holes are then well rammed up with ganister and red sand from the outside against an inside support, there being a round bar inserted to form the tapping hole, which is not more than 2 or 3 inches long. The fire is then lighted and the breast built up with hard burned coke, the plate put on and black sand rammed between, the tapping hole in this plate being made with ganister. After the blast has been turned on, it is allowed to blow through the tapping holes for twenty minutes or half an hour in order to get them thoroughly dry, which is important, otherwise much trouble may be caused, perhaps necessitating the use of a sledge bar. About the average time of starting the fire is two hours before the iron is charged, and the coke should be from 12 to 18 inches thick above the tuyeres, burning well and evenly throughout, but it is possible to tap out in an hour from lighting up. Each time after tapping, the hole is stopped with a mixture of clay and coal dust, the latter preventing the clay from setting too hard and causing it to crumble when tapping out. The stopping is applied on a rod, and is best inserted at a slight angle from the top of the metal, which lessens the risk of it being washed away and the consequent filling of shank ladles too full, running upon the floor, and general confusion.

It is almost impossible to work a cupola to a time-sheet, as one day it may be wanted to work fast, or perhaps two may be on together; another day cold or hot, hard or soft, according to the quality of castings required, that is generally speaking of contract shops; but with railway work, which is so regular and well known, it is very much easier to regulate and approach somewhat time-sheet working. Foundries noted for turning out castings of the required grade, run their heats separately, which can be done without dropping the bottom of the cupola, by clearing out through the breast the clinkers, &c., until it is certain that the cupola is clean. This of course takes from half an hour to one hour extra, and more coke is consumed. It appears almost impossible to run two heats of different grades of iron without getting them intermixed. This is especially the case where the charges are small, as there must always be globules of metal remaining about the fuel and sides; however, as a rule, one class of metal is run off before the other is seen to be coming down; especially in the case of cylinders, when the bottom is run quite dry before the cylinder metal is seen to be melting. It is a good practice to put a charge of general iron in between a hard and soft grade. The iron never gets into the tuyeres or blast chamber, as by knowledge of the cupola and a careful watch this is prevented. The slag is also watched through the tuyere holes, and when it is seen to have risen high it is tapped out, perhaps about three times a day. Of course it can be higher than the tuyeres without entering, as the pressure of the blast carries it to the centre of the cupola. The blast chokes and chills the fuel at its entrance, and as soon as this is noted to be of any extent, it is rabbled into the centre of the cupola. Limestone is used as a flux. About 100 to 150 lbs. is sufficient for 4 or 5 tons of metal, a portion being put on with each charge. It also helps to clean the cupola and make it drop better, which is done after the last metal has been tapped out, the cinder and dirt being raked away. The drop is not absolutely the best form of cupola, although very convenient, and is only tolerated in foundries where it is not absolutely necessary to have the first and second heat hot. The bottom, having to be made every morning, is cold and damp, whereas with a permanent bottom it is always dry. In the latter case, the cinder and dirt have to be raked through the breast, and consequently requires more room. It is very rarely that there is any metal left in the bottom, if so, never more than half a hundred-weight, which is picked out of the cinder and is good scrap.

Iron mixings generally consist of one of soft pig and two of good scrap, or two of soft pig and three of good scrap, but if a very soft casting is required, select two brands of pig and no scrap, because every time iron is melted it becomes harder; but it is not always advisible to use all pig, as a certain percentage of good scrap will make the castings stronger and cleaner. One-fifth of good scrap is used in casting the cylinders, the brands of pig being selected from experience: say Madeley Wood, Carron and best cylinder scrap. Wrought scrap and white iron generally give trouble, through not being intimately mixed, unless the charge is large and sufficient time is given for them to blend in the boshes, which will hold at least twice as much before twelve o'clock as after.

The moulding and casting of the double cylinders is about the most important and difficult piece of work in the whole foundry. The casting must be perfectly sound and the metal of high quality, hard and tough, but at the same time capable of easy machining. The metal is poured into the mould as hot as it is practicable, because the hotter it is poured, the sounder and cleaner the cylinders will finish. Smooth-skinned cylinders may be obtained by dull casting, but soundness is sacrificed for smoothness. Hot pouring is a great trial to the mould, because the metal will find its way into poor joints and vents, the latter causing the casting to blow and scab, also destroying cores or mould if burnt in drying. There is not much risk as a rule in the mould; but the danger lies chiefly in the cores, as they are numerous and of complicated shapes. Great attention is bestowed in securing the vents and setting the cores in their prints, the moulder clearly satisfying himself that the loam does not plug up vents, and as little is used as possible, especially if the mould is going to be cast immediately. The hay rope round the body cores is bound as tight as the rope will stand, or it will slacken, and then it will be impossible to sweep up the cores solid and true. In making the cores a finer sand is used for the face than is required for the centre, where cinders are often used. Fine sand gives a smooth surface, but is bad for getting the gases away. They are put into the drying stove as soon as possible after making, because air-dried cores crumble. The cores are either made in boxes, struck up from a board, or as the S cores, partly made in a box and partly swept up, which is very handy for the moulder and easy for the patternmaker. The shape of the core-boxes can be clearly surmised by reference to the cores in the drawings of the cylinders, Figs. 57–60. The patterns are shown in plan and elevation in Figs. 53–56, all core prints being indicated by diagonal shading.

When the cylinders are not moulded from iron patterns on a machine, the following method may be taken as an example:—The top half or steam chest portion is placed at the bottom, in order to get all the steam passages and valve faces perfectly clean. Place the joint of the pattern K, Figs. 53–56, upon a "turnover board," which in this case is an iron bed-plate on the floor level, and line up with facing sand; put on the middle box, and ram up with backing sand. Vent all round with $\frac{5}{16}$-inch or $\frac{3}{8}$-inch vent wire, placing $\frac{1}{2}$-inch iron in front of the leaders L, Figs. 58 and 60, to strengthen and prevent the metal breaking into the mould. Gaggers, like loose irons, are placed between the body core prints, to bind the mould at this point; also between each of the webs M, Fig. 53, the valve rod core, and in front of the blast pipe seating N, Figs. 53 and 58, N', Fig. 53, being loose formers, which remain in the mould when the pattern is drawn, and afterwards removed. A shallow groove is also formed and filled with cinders at M, Figs. 53, 57, 58, and a

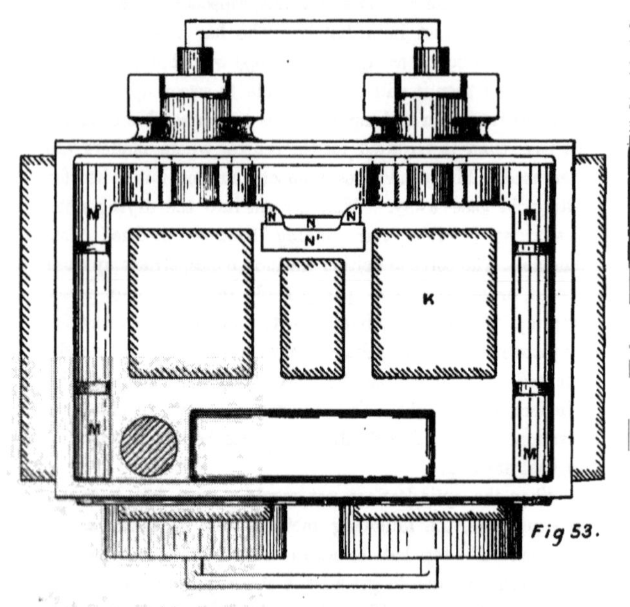

Fig 53.

⅜-inch vent wire is pushed down to the foot of each gagger. An angle-iron is used sometimes at this joint to strengthen it. Parting sand is then sprinkled upon the core prints and joint, the bottom box placed on, rammed up and vented all over, no gaggers being required in this box. This portion of the mould is then turned over, placed in the pit, and the joint made good and solid; that is, although it has been rammed from the opposite side, it is tested all over to find the weak places. The other half of the pattern P, Figs. 54–56, is then placed in its dowels, lined up with facing sand, the top box put on, gaggered, rammed up with backing sand, and vented all over as indicated in Fig. 57. The top box is then lifted,

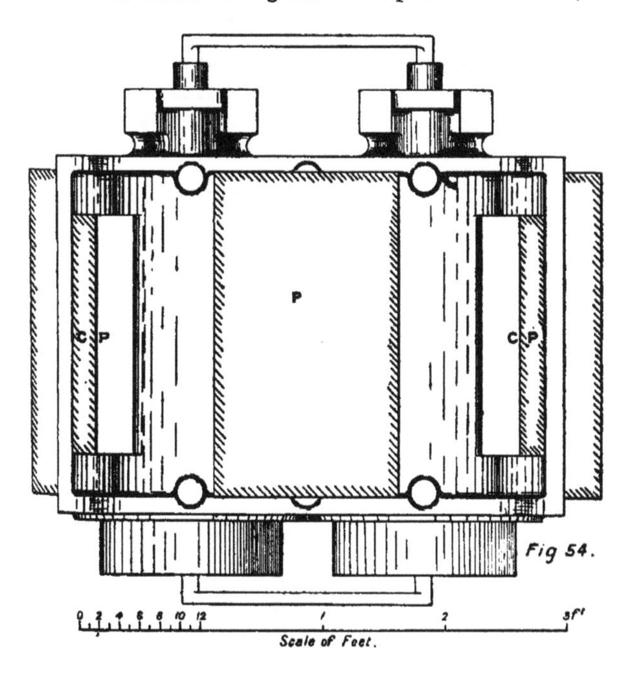

Fig 54.

turned over, and the pattern drawn, using the crane straight away to separate the middle from the bottom, drawing the pattern from the middle. When the middle box is separated from the bottom it leaves the steam chest core prints, the core N, Fig. 58, and the loose formers N', Fig. 53, so that each portion of the mould can be easily got at and finished, making good all portions disturbed by drawing the pattern, and then fin, as little as possible consistent with a risk of

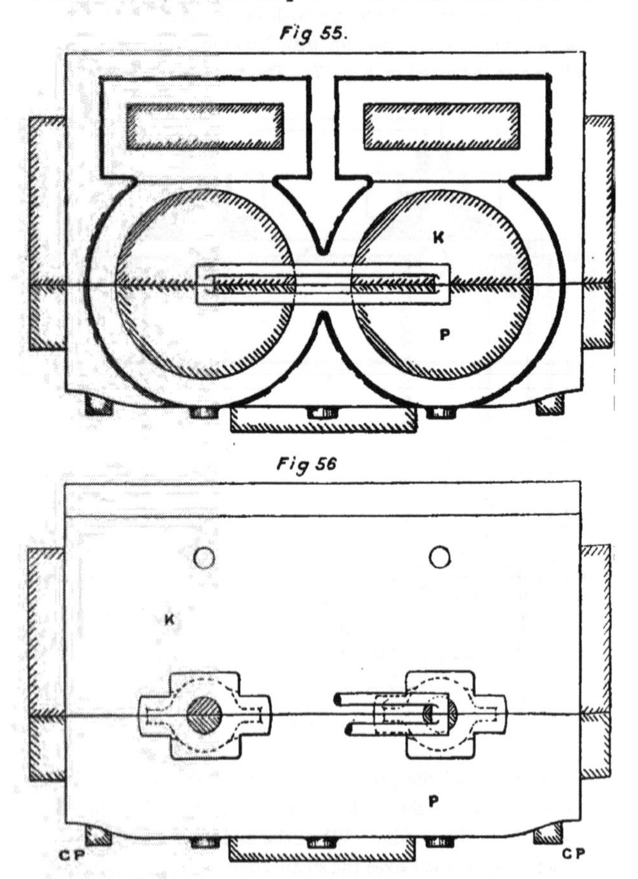

crushing, as a fin is always preferable to a crush. Sleak over with blacking, which gives a gas-evolving, or kind of spheroidal coat, and protects the metal from the chilling effect of the sand. Place the middle box upon the bottom, and fix the two side cores, Figs. 57 and 60, by the bolts *b*. The core irons, vents and "rickets," by means of which the gases strike away through the bolt holes *b* are clearly shown,

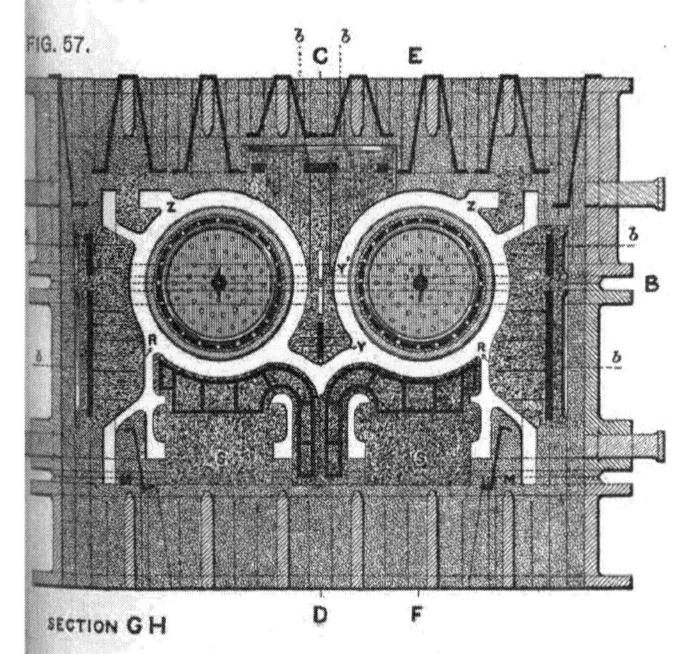

and this remark will suffice for all side and lightening cores. The moulds are then well dried; afterwards the bottom portion is separated, and the steam chest cores fixed, S, Figs. 57 and 59, making good all round the joints of the prints, these cores being held down by bolts to the bottom of the box. The drain chamber core is then fixed upon four studs

T, Fig. 58, which by means of two 1⅛-inch round cores, form a communication between the two steam chests, and further by means of the cores T', Figs. 58 and 60, with the pet cocks. The S and exhaust cores are then fixed and all joints made

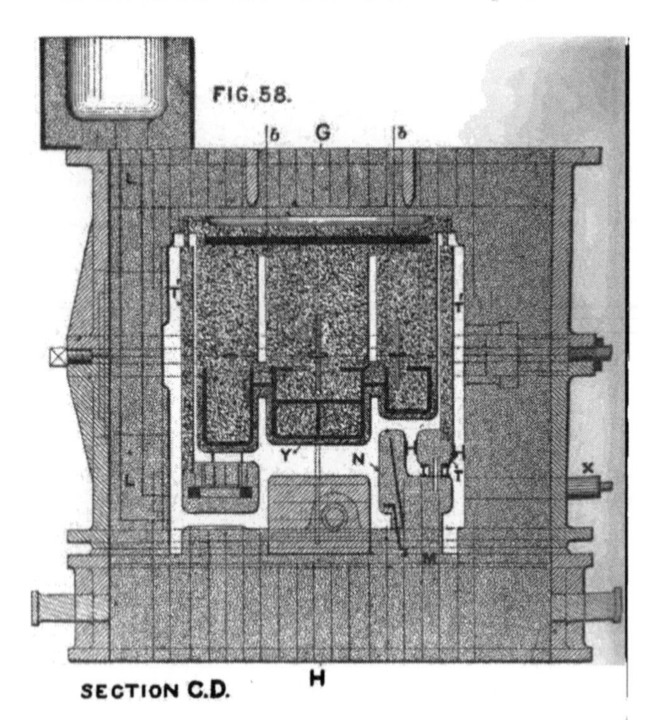

SECTION C.D.

secure; Figs. 57 and 59 showing them in position secured by the bolts *b*, the vents in the S cores being formed by a straight wire and joined at the curves by a piece of string, the gases striking through the holding-down bolts. Place the steam chest cover core in its recess V, Fig. 59, and drop the middle box upon the bottom, then carry the core V^1, Fig. 59, forward, making the back good with sand, fix in the vent pipe W, Fig. 59, which enters the cinders of the steam chest core S.

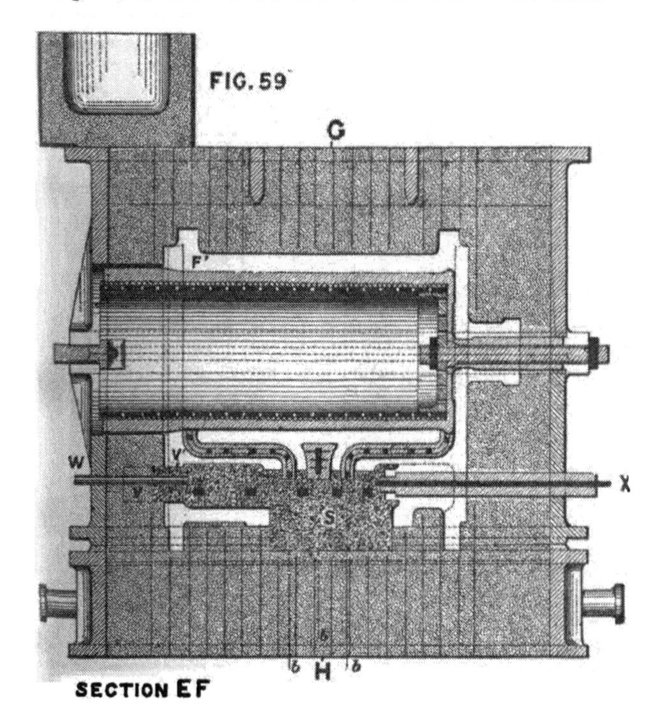

Then well stop up the joint between the two boxes with loam, and place the valve spindle cores X, Figs. 58–9, the prints for which have been formed by pocket cores, and loam the joints of these up. This portion is then placed in the drying stove again for twelve hours and afterwards placed in the pit, the sand of which has been lightened up and sprinkled with hay. The centre shaping core Y, Figs. 57, 58, 60, is then placed upon studs. There is generally about $\frac{3}{16}$ inch space between

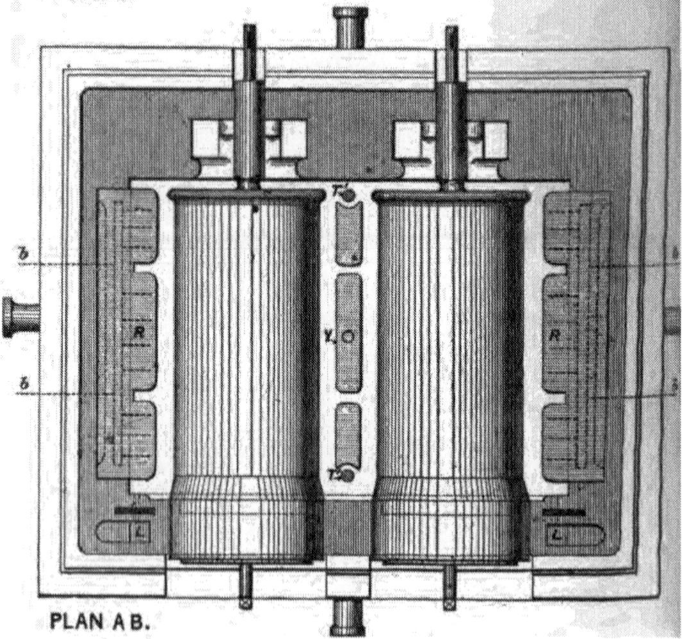

FIG. 60. PLAN A B.

this core and the top shaping core, which is made up with loam to prevent crushing, the gases from this core being conveyed by the $\frac{3}{8}$-inch vent holes Y′, Figs. 57, 58 and 60, from the cinders in the bottom core to those in the top. The two body cores are then fixed, and the whole passed by the foreman. The boxes and body cores are then cottered up, and joints made secure with loam, especially at the fin F, Fig. 59. It will be observed that there are not any feeders or risers beyond the leader, and these have not been found to be necessary, the head of metal in the basin over the leaders being quite sufficient. By means of the core prints C P, Figs. 54 and 56, and the core Z, Fig. 57, these patterns can be utilised for another class of engines, having a leading bogie.

Fig 61.

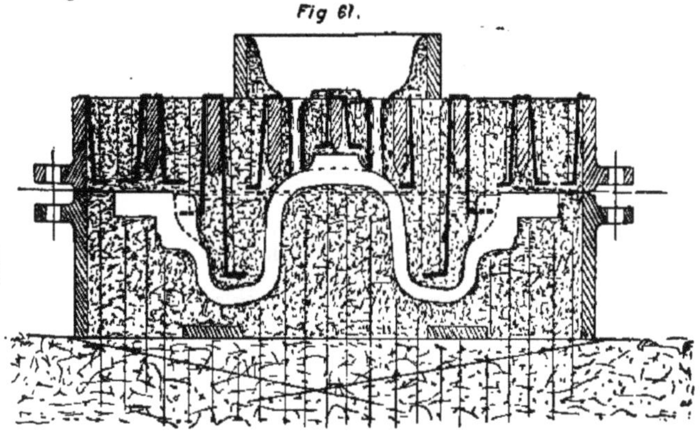

Scale $1\frac{1}{2}''$ = 1 *Foot.*

Fig. 61 shows the method of casting the covers, and Figs, 62 and 63 the front and top steam chest covers, the two latter being examples of plate moulding, the core in Fig. 62 being shown in dotted lines. Fig. 64 is the piston head, and when a moulding machine is not used, Fig. 65 shows the method of casting barrels for the piston rings in the foundry floor.

As the cylinders may be taken as the best examples of setting cores and securing their vents, the foot or drag plate may be taken as an example of an irregular joint, and the setting and wedging of chaplets. It consists of a heavy

Fig 62.

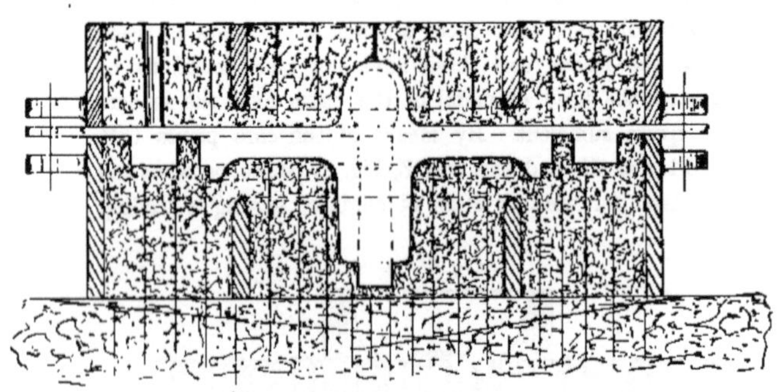

Scale 1½" = 1 Foot.

Fig 63.

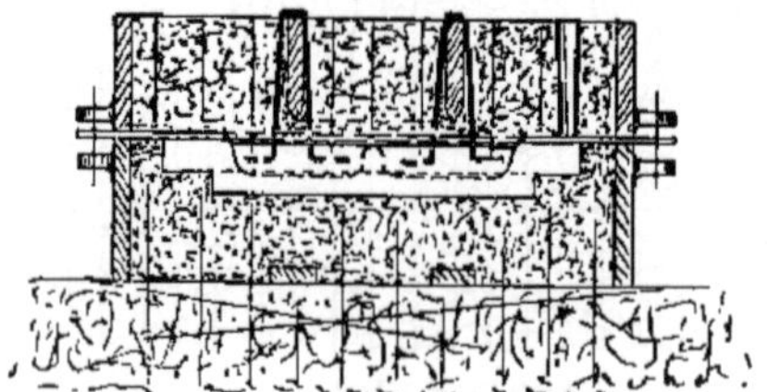

Fig 64.

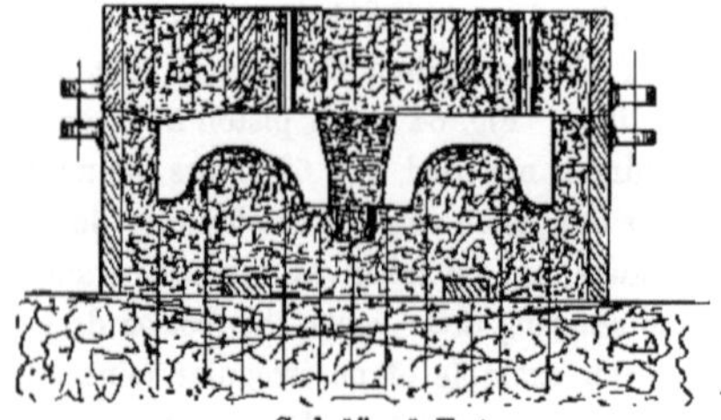

Scale 1" = 1 Foot.

casting, having projections for securing the brake cylinder and the brake shaft carriers. It is shown in longitudinal section in Fig. 66, and transverse section in two planes in Fig. 67, Fig. 68 with its scale being an isometric view of the core M, which is the principal one in this casting, the minor cores being the lightening ones M'. The top halves of each are necessary as making-up pieces, because the prints of these lightening cores and the body core M are nearly at the bottom of the pattern, which has to be drawn, and N, Figs. 66 and 67, is secured by the bolts *b* for the clearance

Fig 65.

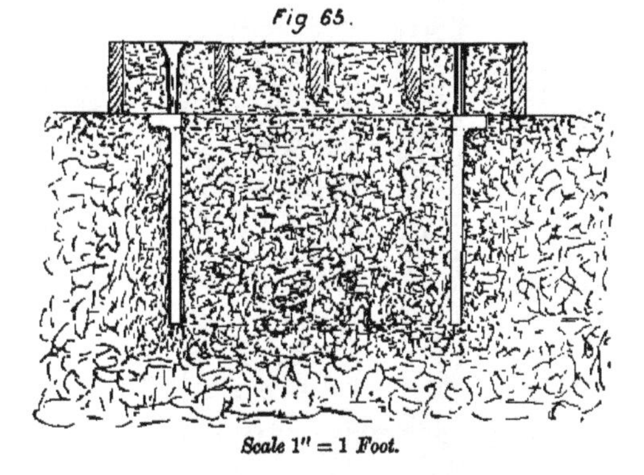

Scale 1" = 1 *Foot.*

space, for bolting the casting to the frames. It is bottom poured at P, the tail end of the casting, the leaders having a break and connecting runners for convenience, which also lessens the fall of the metal, the facing strip P' being formed by a loose strip on the pattern, and there is a riser at R, Fig. 66.

In the first instance the pattern is fixed face downwards and packed up, the bottom box placed in position and rammed, cinders being put in layers opposite the holes in the side of the box for venting. After the first casting has been

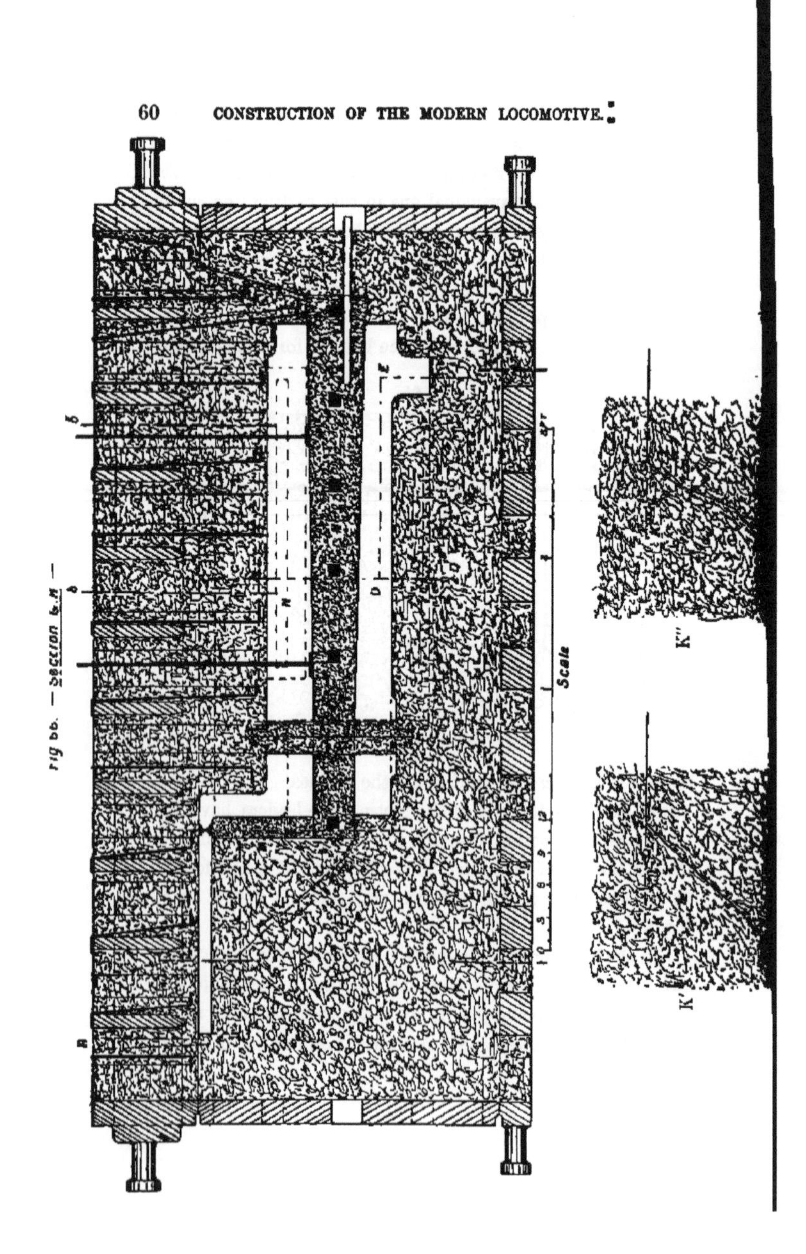

Fig 55. — Section b.a —

made, the top joint is used as a turnover board for the pattern, which acts as a kind of print and indicates at once the exact position the pattern should occupy, without any further measurement or setting. After the bottom box has been rammed up it is turned over, and if the top box has been used as indicated, the old sand is knocked out, the box placed in position, the pattern lined up with facing sand, gaggered about every 4-inch space, rammed with backing sand, and vented all over.

Reference has already been made in the general remarks to the making of irregular-parted joints. As a rule, an angle of about 60° to 80° will give a satisfactory lift, as at K, Fig. 66. K′ would give an easy free lift, but necessitates using a great number of gaggers, whereas K″ would only be made use of by a most inexperienced moulder, and such a man would use a riddle full of sprigs in making this joint. The core M does not require bottom chaplets, because the core iron is of sufficient strength to carry it from print to print, such a core iron being capable of carrying a core of 6-feet span between prints. This core is vented by the gases striking through the cinders, and conveyed by the vent pipe through the side of the box. Another method, and the one most frequently adopted in venting this core, is to bring the cinders up to the joint to meet those in the box, and then the joint is securely made, so that the metal cannot burst in on the cinders. The core for the drag pin is placed in position, and the joint passing through the large core M secured.

In the section, Fig. 66, only two chaplets are shown, but five are used; three along the centre, and one each side, at the broad end. They must be so wedged down that the pressure of the metal due to head cannot force the core up and destroy the formation of the casting. Their heads must be made to rest fair and square upon the core, the moulder

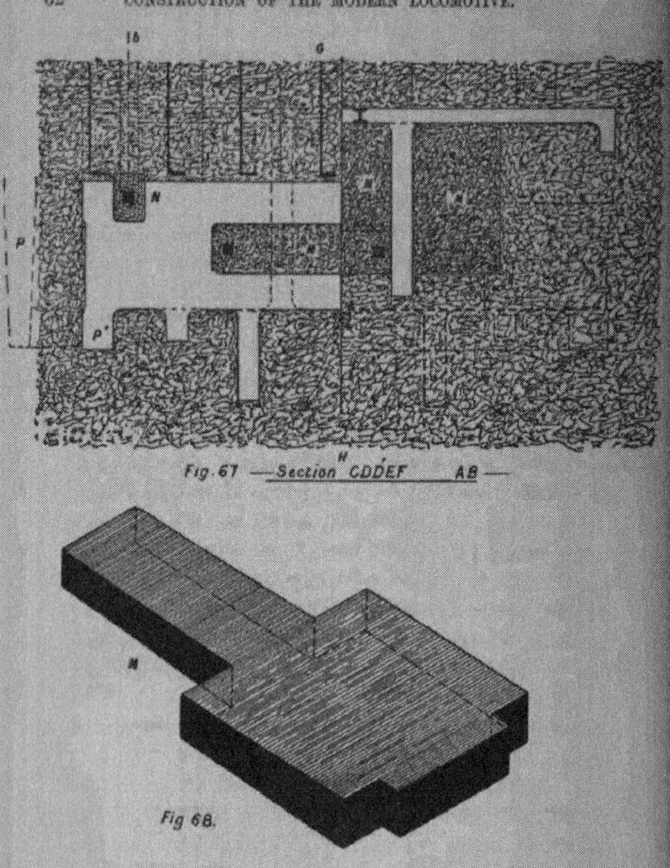

Fig. 67 —Section CDDEF AB—

Fig 68.

clearly satisfying himself by trial that this is the case, and as little blocking (which should be avoided) and wedging as possible. All this is absolutely necessary, because frequently castings are lost through inattention to chaplets. In the case before us, it is the custom to place a 5-ton weight upon the top box for weighting purposes, and it is so fixed that there shall not be more than about $\frac{3}{4}$ inch space between the bottom of the weight and the top of the chaplet stem. It is then possible to fairly secure them with two small wooden wedges. There should be a small clearance of sand round the chaplet stems, otherwise in pushing them down upon the core, or the vibration set up in wedging, may cause sand to fall out and damage the contour of the casting, the metal forming small fins round the stems. In some cases buttons are formed in the top joint, so that if the metal blows round the stems, it can be chipped away in dressing to form a good surface. When bottom chaplets are necessary, equal care must be bestowed upon setting them, especially if wood blocks are used. The blocks must be hard, so that the chaplets will not settle with the weight of the core or the wedging of the top chaplets. Such large moulds as the foot or drag plate are only skin dried, and cast while hot, which is necessary to prevent moisture striking through the skin again. This gives a good surface, and at the same time the backing sand, not having been baked hard, is very porous, and allows the gases to permeate with great freedom.

Fig. 69 is a longitudinal section of the combined sand box and splasher, Fig. 70 a transverse section through two planes of the sand-box, and Fig. 71 a simple section through the splasher, $\frac{5}{16}$ inch metal throughout, except at the foot and stud holes. Six core-boxes are required, viz.: sand-box, inlet and outlet for the sand, bridge for spanning over the inlet core, foot of the sand-box and splasher core. The sand-box core is vented by means of the pipe, the gases passing up the channel scooped out of the sand, Figs. 69 and 70, and the splasher core by means of the vent wire, the gases striking through the print, Fig. 71. Studs and chaplets are carefully

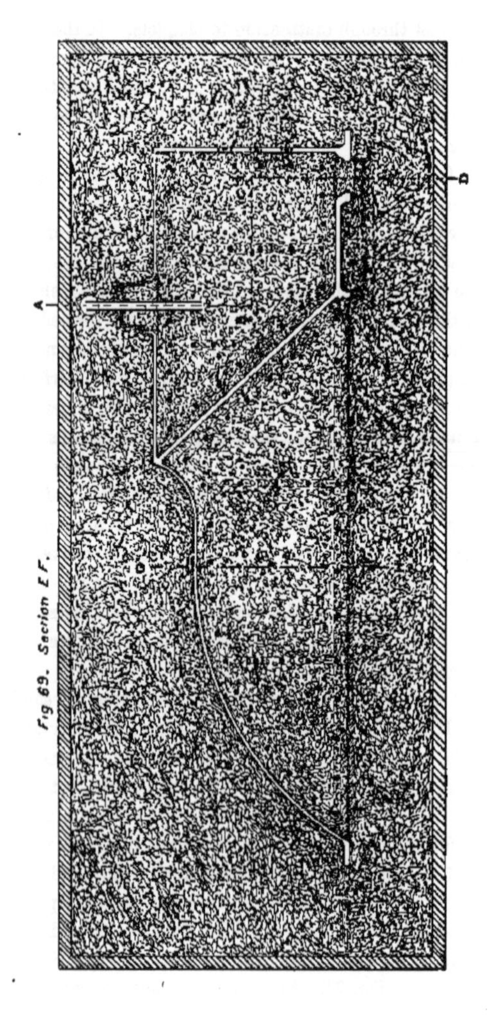

Fig 69. Section E F.

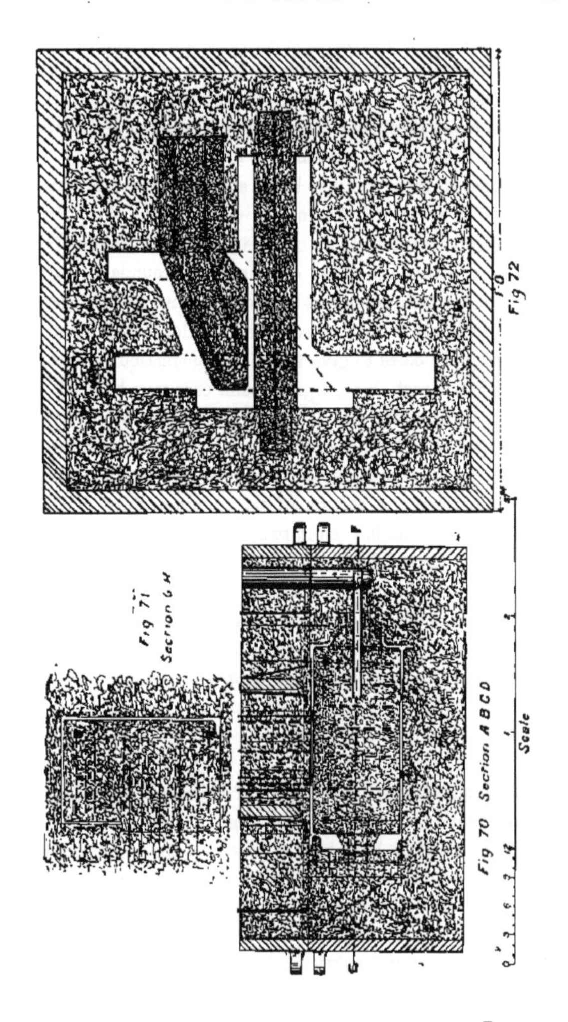
Fig 72
Fig 71
Section G H
Fig 70 Section A B C D
Scale

set and wedged in the position shown, and the casting is run by means of three gates 2 inches by $\frac{1}{2}$ inch at K, Figs. 69 and 70, upon the top of the front edge of the sand-box, not seen in the views given, but the position of which is indicated by the dotted lines, a riser coming off at R, Fig. 69.

The metal in these castings being so thin, great care must be exercised or many will be lost in contraction. For this reason the cores contain more cinders than usual, especially the sand-box, where the metal completely surrounds it, and the core irons, which are generally covered with loam to the depth of 1 inch, are in this case covered to a depth of 2 inches, thus enabling the cores to give with the contraction. When the splashers are cast alone, a green sand core may be used.

Fig. 72 is a section through the joint of the valve-box, which is attached to the above sand-box. It is run by means of a $\frac{3}{4}$-inch gate on one flange and a $\frac{5}{8}$-inch riser on the other. Fig. 73 is a transverse section of the chimney through the smoke-box. It is moulded from two iron patterns, that from the seating on the smoke-box to the core joint forming one portion, and from the core joint to the top the other. The bottom core print of the latter enters a socket, which is part of the bottom box. The first operation is to line up this socket with facing sand, and then drop the bottom portion over, firmly bedding it down, and ramming it up on the inside. The second box is then placed in position, and the pattern for the chimney fixed in. The gaggers are then put in this box and rammed up, a layer of cinders being placed at the joint with the third box, which is not broken after the third and fourth boxes have been rammed up, the joint proper being at the top of the coping, which has to be broken to draw the pattern. There is one circle of $\frac{3}{8}$-inch vents round the top, which meet the needle vents under the coping. The core is struck up, and consists of one round of hay band tightly bound to the barrel, with one thin layer of clay to stick the loose straws down, and one coating of loam. It is then thoroughly dried. Afterwards the finishing coat of loam is put on and the recess at the coping made. The core barrel rests upon the iron of the core print at B to prevent crushing,

Fig. 73.

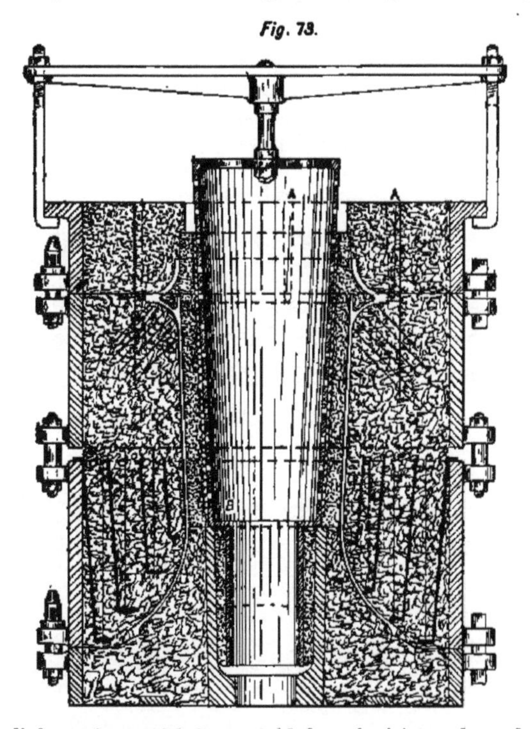

a little parting sand being sprinkled on the joint to keep the metal from bursting in. The casting is poured from the two points A, the metal entering the mould by three branches from each gate, a riser being upon the opposite side.

The moulding of the blast or exhaust pipe is shown in

section, Fig. 74, and in plan, Fig. 75. This method is adopted in cases where a foundry is short of moulding boxes. To use a bottom as well as a top box is the better plan, as it facilitates the work. In this case the foundry floor must be prepared, the pattern firmly bedded in, and the position of the top box secured by stakes or angle irons. The casting is run by means of the inch gate C, and an inch riser R. The necessity

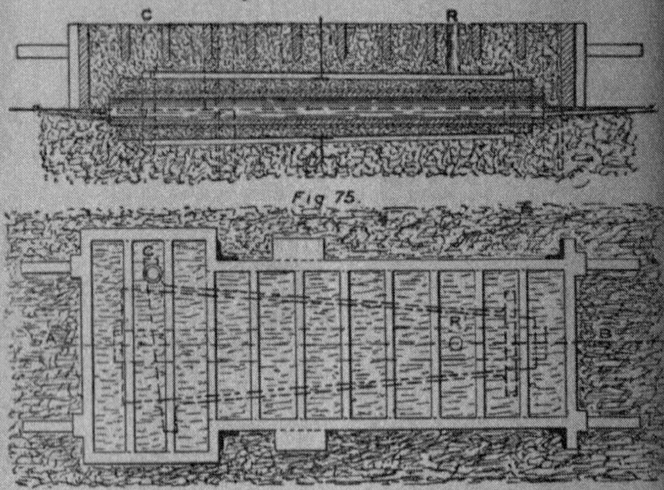

Fig 74. Section AB.

Fig 75.

for venting is not very great, the body of metal being small, the only provision being for the air to strike away from the core as shown by the arrows, Fig. 74.

In Figs. 76, 77 and 78, is clearly shown the method of casting the fire-bars on a plate pattern, the angle irons at the corners of the box being guides for the plate. It is a very quick and trustworthy method, the wasters being extremely few. In Fig. 79 is seen a half-size transverse section of the plate pattern, showing the bar cast on flat with a slight inclination, by means of which whatever small quantity of dirt

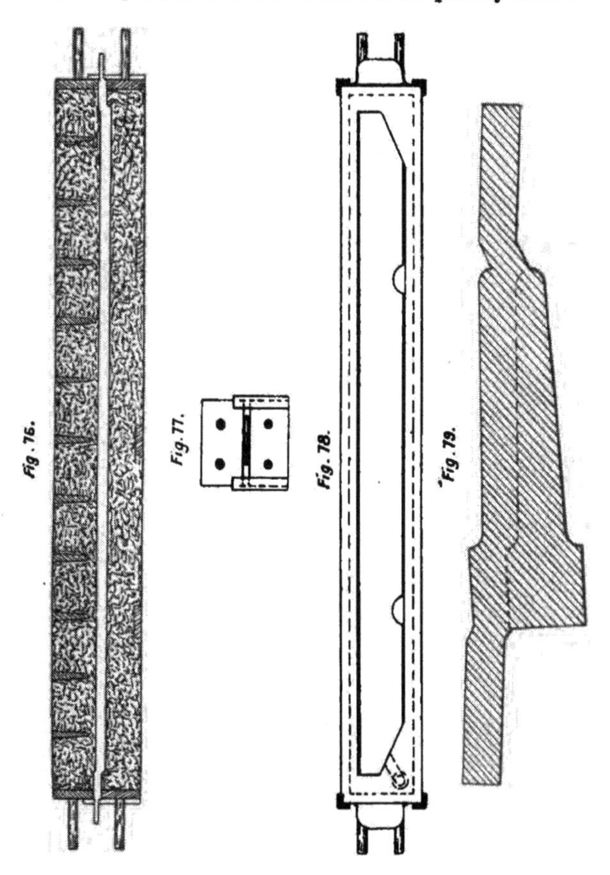

goes over with the metal, rises to the least important part of the bar. This also causes the least amount of sand to be lifted in the top box, and the least risk of disturbing the mould when drawing the pattern. Small bars may be cast vertically, and three or four on one plate.

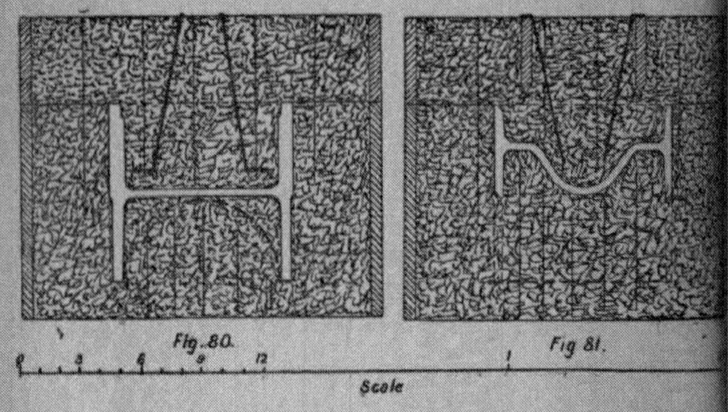

Fig. 80. Fig. 81.

Figs. 80 and 81 are sections of the axle-box keep, the box itself being either a steel or a brass casting. They are top

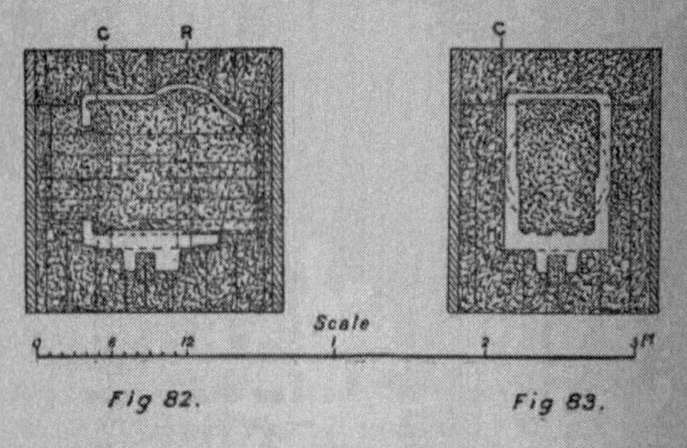

Fig 82. Fig 83.

poured, the metal entering the mould by two branches from one gate, a riser being upon the opposite side.

Figs. 82 and 83 are sections through the tender axle-box.

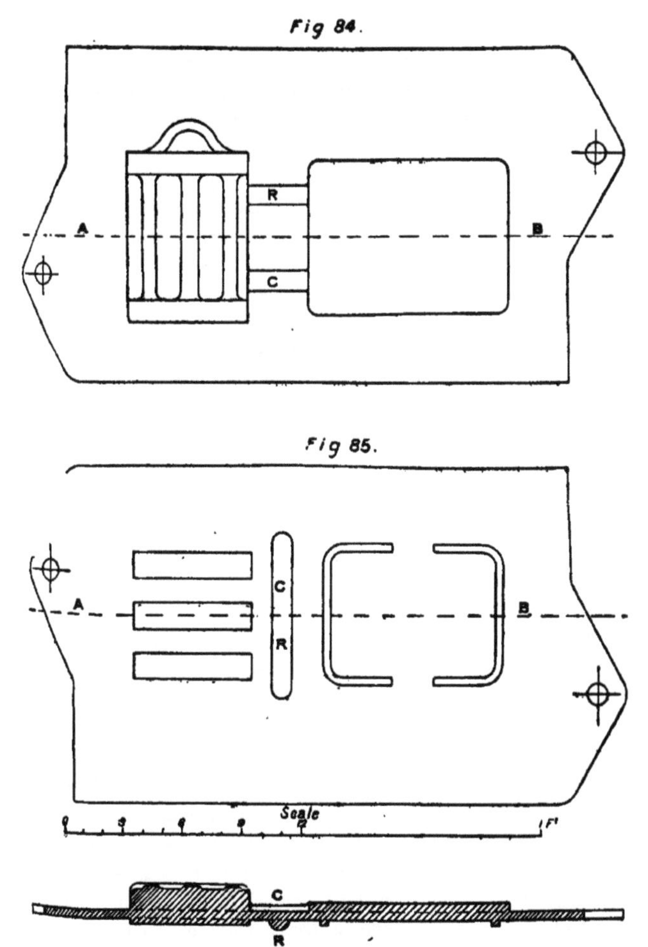

Fig 86. Section AB.

They are cast by means of a $\frac{3}{4}$-inch gate at C, to one of the flanges or horn-block guides, and a $\frac{5}{8}$-inch riser at R. The core is bound together by loose irons, as distinguished from a regular core iron, and straight vented with needle vents, the mould being also vented at the sides and bottom. Figs. 84, 85 and 86 are the plate patterns for the front lid of the tender axle-box, and the packing or making-up piece between the top of the box and the brass, C R being connecting runners or gates which are formed when the pattern is made, by the moulder scooping out the sand with his trowel.

As nearly one hundred patterns are required for iron castings for a locomotive, besides steel and brass, it can easily be surmised that these illustrations might be carried on to a greater extent, but the author considers the examples given as fairly representative of iron foundry practice, and to carry them further would simply be a protracted use of the terms "line up with facing sand," "ram" and "vent."

Part II.—The Use of Steel Castings.

The introduction of trustworthy mild steel castings has replaced locomotive iron castings and forgings to the extent of about forty articles, and is still increasing. These are illustrated by Figs. 87–108, including stays, gusset or otherwise; horn-blocks and axle-boxes, domes, covers, and safety valve seatings; spring links, brackets, and fire-box foundation rings; reversing shaft—Joy gear—motion plate and brake shaft; wheel centres, and, to a limited extent, the unhammered crank axles made by Messrs. J. Spencer & Son, Newburn Steel Works; one of which, supplied to the North British Railway for engine 460, a heavy class of goods engine working trains over the main lines, has run a mileage of

337,879 miles from February 19, 1881, to April 27, 1892.* This would of course be considered a good mileage for a mild steel forged axle, and is mentioned as an interesting fact; but under existing circumstances the author does not deliberately advocate the use of steel castings for crank axles, although the experiment deserves encomium, and their adoption may be in the future an accomplished fact.

Advantages may be gained in many points by the adoption of steel castings, because they can be guaranteed to be sound, homogeneous, and free from internal strains, at the same time possessing the specified tenacity and ductility. They contrast favourably with built-up iron forgings and those from mild steel ingots; remarkable results as to ductility having come under the personal experience of the author. The tensile strength of unforged steel castings may be taken at about three to four times that of cast iron, therefore they may be considerably lighter than those of cast iron of equal strength; but whenever steel castings have to replace forgings or iron castings, the physical properties expected of those castings must be carefully taken into consideration, ductility being of much greater value than an unusually high tenacity, because it will give a greater amount of endurance and furnish ample warning of failures, excepting perhaps in a few cases, such as pins, where a great amount of wear and tear takes place, due to friction.

Steel castings are designed sometimes of a peculiar shape and thin in section; when this is the case, the steel founders have to contend with several difficulties in manufacture, which could be avoided if the designer would keep in view the fact that steel castings should be as simple in design and as near uniformity in thickness as the circumstances of the

* In April 1894, Mr. Holmes, the chief mechanical engineer, informed the author that this axle was thoroughly examined and tested during the last week in March, was quite satisfactory, and is still running.

case will admit, and of easy curves. It is contended by some American engineers that it is preferable to bolt several small castings to one large one, than to make a complicated casting with several branches, brackets and projections ; because each change in the form of the casting is a source of uncertainty. Steel wheel centres having the rim, boss, arms and balance weight formed in one casting, compare very favourably with built-up wrought iron centres, in which there are of necessity so many welds. To overcome the manufacturing difficulties, it is of the utmost importance that the rim should be in proportion to the arms, which are mostly of the oval section, the dimensions of the major and minor axes where they enter the boss being $4\frac{1}{4}$ inches by 2 inches, and at the rim $3\frac{3}{8}$ inches by $1\frac{5}{8}$ inches, Figs. 93 and 94. Mr. R. A. Hadfield opines that the most trustworthy job is made of wheel centres by having a small strengthening rib on the inside of the rim and another at the boss into which the arms sink, which is chipped out during fettling; also that in the case of crank boss wheels, trouble is sometimes experienced when the counterbalance weight is heavy and relatively out of proportion to the other parts of the wheel. This large mass of material near the outside cools more slowly than the rest of the metal, thereby tending to draw it out of shape. Therefore, to obtain a thoroughly sound and good steel casting, the balance weight should be distributed over a sufficient number of spaces between the spokes. The adoption of steel roof-bars for fire-boxes has proved a success see Fig. 15, p. 13. It enables the stays to be well proportioned, and convenient attachments for any form of sling stays desired can be easily added. With cast steel hornblocks and axle-boxes there is greater strength without being cumbersome, greater security in running, and the risk of breakage is reduced to a minimum. Cast steel motion-plates give greater stability as frame stays. Being cast in one piece, there are not any rivets to work loose, as in the case of the built-up type, and, having greater strength and elasticity than cast iron, they can be made lighter, and there is not the same danger of breakage from sudden shocks. Generally, the remarks made as to the design of iron castings apply themselves doubly to steel castings, especially those relating to fillets, abrupt changes, crystallisation, and the internal strains set up by unequal contraction.

The process and the raw materials should exercise a great influence upon the choice of steel castings, because they can be produced short and brittle, which no amount of subsequent annealing will rectify, for reasons well known to metallurgists. Up to the present, as far as the author is aware, the acid process has given the most uniformly successful castings, the basic and the Bessemer generally favour oxidation of the metal during the process, and this incorporation of the oxides of iron is particularly obnoxious, not only in steel for castings, but generally, although upon occasions splendid castings have been made by these two latter processes, and at the same time containing a low percentage content of carbon. The enormous temperature required to produce fluidity, the difficulty in obtaining a suitable facing sand sufficiently refractory to withstand this great temperature, and at the same time porous enough to allow the gases to permeate, besides the direct venting and the mould being to a certain extent fragile enough to allow the greater contraction without pulling the casting, all tend to produce a product which cannot compete, as far as surface is concerned, with iron castings.

A peculiarity relating specially to steel castings, is the great diversity of opinion and want of uniformity in specifications. Different engineers have different ideas both as regards the quality required for certain purposes and the size of the test bars. The latter vary from 2 inches to 10 inches long, whilst an elongation of 20 per cent. to 10 per cent. is specified for the different lengths of 2 inches, 5 inches and 8 inches long. The same remarks relate also to the bending test bars. Different lengths, round and square in section, and the angle of deflection which the bars must sustain without fracture, are specified. Another difficulty is in nominating the grade of steel required. For instance, there should be a distinct difference in the specified grade for the steel for a gear wheel and that of a locomotive driving wheel. In the former case a material is required to withstand a considerable amount of wear and tear, whereas in the latter the wear and tear takes place in the tire, and the wheel centre is required to be tough to withstand all the strains set up in passing over rough roads, crossings and round curves. Due consideration should be given to the shape, and whether the required product is to be in tension or compression. It would therefore be conducive to general excellence, and facilitate the progress of steel castings to a desired end, to formulate a standard specification for the grade of material required for locomotive work, and recognise a standard size for both the tensile and bending test bars. This would be certainly advantageous to both consumers and founders.

At the Newburn Steel Works of Messrs. Spencer & Sons, mentioned previously as the makers of unhammered crank axles, four different grades of material are manufactured, viz.:—(1) An extra soft steel of 28 tons to 32 tons per square inch ultimate tensile strength, and 25 per cent. to 15 per cent. elongation, which is applicable only to castings of fairly plain section, for all purposes when toughness is essential, such as wheel centres, cranks — marine or otherwise—cross-heads, motion-plates, horn-blocks, bogie frames, and locomotive work generally. This material would give a minimum elongation of 15 per cent. on 2 inches, and would be capable of bending while cold through an angle of from 90° to 60° respectively. (2) A medium steel of 32 tons to 35 tons per square inch ultimate tensile strength, and 15 per cent. to 8 per cent. elongation, suitable for heavy gearing, boiler seatings, &c. (3) A medium hard steel of 35 tons to 40 tons per square inch ultimate tensile strength, and from 8 per cent. to 3 per cent. elongation, suitable for all kinds of gearing, hydraulic cylinders, permanent way castings, tram wheels, and all castings where a maximum wearing capacity is required; and (4) a hard steel adapted for special purposes requiring great hardness and strength.

The annealing of steel castings is a matter of great importance, especially when toughness and ductility are essential. By this process the internal strains set up during casting in the different parts have time whilst at a uniform temperature, and when the crystalline structure is changed into an amorphous or plastic state, to adjust themselves to one another, and, according to Sir William Siemens, it must therefore be most disadvantageous to allow the castings to cool right down, whereby these differences of strain would be brought to a maximum, and very likely cause an evil that was preventable. In order to anneal a casting to the greatest advantage, it should be removed from the mould to the annealing furnace without allowing it to cool; but no doubt a considerable amount of judgment must be exercised in removing a casting hot from the mould, and it is quite sufficient to specify that each casting must be thoroughly annealed, and allowed to cool gradually in such a manner that there shall be no undue strain in any part.

Table I. gives the results of some experiments on the tensile strength of specimens taken from ordinary castings, without any subsequent hammering, in order to ascertain how the mechanical properties of steel castings are affected by annealing, and also by its chemical composition. Five groups, each containing four specimens, were tested; the first three pieces in each group being annealed, and the last in each unannealed. The experiments were made at the Newburn Steel Works.

The advantage gained by annealing may be expressed by a percentage. This suggestion was made by Sir J. Whitworth at the Manchester meeting of the Institute of Mechanical Engineers in 1875. He proposed a mode of comparison of metals by taking the sum of their tensile strengths and percentage elongation; but in the case before us, the percentage contraction of area will be added to this sum, and from Table I. it is seen that varying with the grade of material, the advantage of the annealed over the unannealed is from 10 to 28 per cent.

Table II. gives some further results, obtained in a series of experiments conducted by the same authority as Table I., for the purpose of ascertaining how the tensile strength of steel castings is affected by the annealing process.

From Table II. it will be observed that the annealing process causes a considerable reduction of the combined carbon, the carbons being ascertained by the colour test before and after annealing. The figures clearly indicate the fact, that to the extent shown, the combined or hardening carbon has been changed into graphite or amorphous carbon, which latter does not affect the colour test. Under conditions which are not quite understood, a large proportion of the carbon may separate into the graphitic or amorphous condition, and it is this circumstance which is the cause of the noticeable change. The total carbon, when ascertained by the combustion method, remains after annealing exactly what it was before. There is, however, a possibility that the content of carbon in the surface of the casting may be richer, owing to the absorption from the mould, and that the annealing causes a skin of oxide of iron, which would act as a medium for the combustion of some of this carbon in the surface. A similar change has been noticed by manufacturers of mild steel ingots, and iron castings exposed to long continued heat, which appreciably lessens the content of skin carbon, and causes a peculiar phenomenon. The drillings, however, were taken from the body of the casting and all surface ones discarded. It is clearly seen on comparing these tabulated results, that while the tensile resistance of the original area is scarcely affected by the annealing process, the ductility of the material is very considerably increased.

Tables III. and IV. give the results obtained in a series of experiments on the transverse strengths and torsional resistance of unhammered, annealed steel castings, the specimens being cut from crank webs by the same authority as Tables I. and II.

Table V. gives further results, by different authorities and makers, of the physical properties of steel castings.

Table VI. has been supplied by Mr. James Riley, of the Steel Company of Scotland, and represents the quality of castings turned out by that firm. It is complete and needs no explanation, but the elongations on 8 inches deserve mention.

It may therefore be concluded that steel castings can be supplied very much superior to iron castings or forgings, and to rival those forgings made of hammered or wrought steel. It must be remembered that in a crank axle or any other forging, hammering makes a hard skin which never gets any hotter as the forging cools, consequently internal strains are set up, and the centre metal is of little use from a physical point of view, and the cruciform shape taken by a pipe in an ingot is well known.

From the foregoing results, it would be reasonable to specify that the castings shall be made in close-grained steel

TABLE I.

Test mark.	Analysis—per cent.			Section.		Stress.			Fracture.			Remarks.
	Carbon.	Silicon.	Manganese.	Length.	Diameter.	Original area in sq. inches.	First per. set induced. Tons per sq. inch.	Breaking stress. Tons per sq. inch.	Cohesive force.	Contraction of area. Per cent.	Elongation, per cent. on 2-in. length.	
S A 13	·30	·22	·63	1·75	·533	·223	27·0	31·0	55·1	43·8	24	Annealed.
,,	,,	,,	,,	,,	,,	,,		30·4	55·1	43·8	24	Do.
,,	,,	,,	,,	,,	,,	,,	19·2	33·4	56·4	41·0	24	Do.
								31·6		42·8	24	Average.
H A 13	,,	,,	,,	,,	,,	,,		39·6	46·6	26·84	16	Not annealed.
										28·7	=	Percentage advantage.
S A 18	·35	·23	·61	1·75	·533	·223		38·0	56·0	41·0	22·2	Annealed.
,,	,,	,,	,,	,,	,,	,,	19·0	36·0	57·1	37·06	21·5	Do.
,,	,,	,,	,,	,,	,,	,,			Spoiled in	tooling		Do.
								34·5.		39·03	21·8	Average.
H A 18	,,	,,	,,	,,	,,	,,			Spoiled in	tooling		Not annealed.
												Percentage advantage.
S A 48	·50	·41	·66	1·75	·533	·223		44·0	52·9	16·8	12	Annealed.
,,	,,	,,	,,	,,	,,	,,	24	45·2	48·23	6·3	5	Do.
,,	,,	,,	,,	,,	,,	,,		42·2	48·06	12·3	9	Do.
								43·8		11·8	8·6	Average.
H A 48	,,	,,	,,	,,	,,	,,		44·4	46·2	4·13	2	Not annealed.
										27·1	=	Percentage advantage.
S A 60	·77	·46	·67	1·75	·533	·228		39·8	40·34	1·35	1·0	Annealed.
,,	,,	,,	,,	,,	,,	,,		39·0	40·34	3·3	1·9	Do.
,,	,,	,,	,,	,,	,,	,,	32·4	33·6	34·21	1·8	1·5	Do.
								37·4		2·15	1·4	Average.
H A 60	,,	,,	,,	,,	,,	,,		36·4	36·61	0·8	Nil	Not annealed.
										10·2	=	Percentage advantage.
S A 82	·96	·62	·64	1·75	·533	·223		31·0	31·71	2·24	2	Annealed
,,	,,	,,	,,	,,	,,	,,		38·0	38·3	0·80	1	Do.
,,	,,	,,	,,	,,	,,	,,		35·6	36·2	1·80	1	Do.
								34·8		1·61	1·3	Average.
H A 82	,,	,,	,,	,,	,,	,,		37·0	[illegible]	0·8	Nil	Not annealed.

TABLE II.

Test mark.	Analysis.			Section.			Stress.		Fracture.			Elongation in percentage in a length of—			Remarks.	
	Carbon.	Silicon.	Manganese.	Length.	Diameter.	Area.	First per. set. Tons per sq. inch.	Breaking. Tons per sq. inch.	Cohesive force.	Reduced area. Inches.	Contraction of area. Per cent.	5 in.	4 in.	2 in.		
028	·41	·32	·47	5	·745	·4359	15·26	33·79	36·8	·4003	8·16	12	—	19	Not annealed	Taken from the git of a hydraulic riveter casting.
021	·26	—	—	„	·754	·4464	—	32·8	48·17	·2734	38·7	22	—	37	Annealed	
024	·47	·30	·54	„	·741	·4310	15·5	32·1	33·10	·418	2·9	4·16	—	6	Not annealed	Taken from the git of a hydraulic cylinder casting.
025	·35	—	—	„	·710	·3959	13·26	36·6	50·66	·284	28·11	14·6	—	24	Annealed	
151 S	·35	—	—	„	·754	·4464	11·7	32·3	50·27	·2734	38·7	15	—	28	Do.	
027	·42	·33	·49	„	·758	·4512	19·8	24·01	24·80	·4382	2·8	1	1	—	Not annealed	Taken from the git of a hornblock.
020	·42	—	—	„	·756	·4488	11·8	28·4	34·30	·375	15·9	13	—	16	Annealed	
S 25	·47	·33	·47	„	·724	·4166	13·12	30·08	40·36	·3067	23·05	21·9	23·4	28	Do.	Taken from the gits of pinions.
059	·41	—	—	„	·757	·4500	19·54	38·19	43·05	·399	11·28	—	11·5	13	Do	
060	·41	—	—	„	·757	·4500	19·04	37·69	42·37	·400	11·04	—	10	12	Do	
0157	·35	—	—	„	·754	·4465	19·0	38·3	45·60	·375	16·01	14·5	14·5	16	Do.	Taken from the gits of horn-blocks & axle-boxes.
0158	·35	—	—	„	·754	·4465	18·69	38·5	48·35	·3462	22·46	19·25	20	22·5	Do.	
0159	·35	—	—	„	·747	·4382	18·64	38·4	47·59	·3536	19·30	17·8	18	21·5	Do.	
06567	—	—	—	„	·539	·2281	18·60	31·32	59·83	·1194	47·65	—	—	26·5	Do.	4 ft. 4 in. wheel centre.
06568	—	—	—	„	·548	·2358	15·19	30·24	56·77	·1256	46·73	—	—	28·0	Do.	2 ft. 6 in. wheel centre.
06883	—	—	—	„	·540	·2290	17·5	35·6	39·92	·2042	10·8	—	—	14·5	Do.	Trolley wheel.
06915	—	—	—	„	·540	·2290	13·62	28·4	34·10	·1907	16·7	—	—	32·0	Do.	Piston.

TABLE III.

Test mark.	Section.				Stress.		Strain.		Remarks.
	Length of the Bar.	Distance between Supports.	Size of Specimen.	B × D².	Elastic limit.	Maximum Stress in tons per sq. in.	Deflection.	Angle through which the Specimen passed.	
	inches.	inches.	in. sq.		tons per sq. in.	tons per sq. in.		deg.	
Δ 79	14	10	1¼	1·9531	24·00	45·25	5·73	127	Not broken.
„ 84	„	„	„	„	27·77	56·53	3·55	87·5	Broken.
„ 85	„	„	„	„	27·77	49·37	1·12	29·5	Do.
„ 87	„	„	„	„	28·11	54·85	5·80	126	Not broken.
„ 89	„	„	„	„	27·42	51·42	5·50	119·5	Do.
„ 90	„	„	„	„	27·42	51·42	5·40	119	Do.
„ 96	„	„	„	„	28·11	51·43	5·84	125	Do.
„ 97	„	„	„	„	27·42	51·77	5·71	124	Do.

of uniform quality, perfectly sound, free from honeycomb or other defects, and of 28 to 34 tons tenacity per square inch, with a minimum elongation of 15 per cent. on 2 inches; longer bars being difficult to obtain sound when attached to castings. The bending tests to be made upon bars 1¼ inches square, which should be capable of bending cold without fracture over a radius not greater than about one and a half times the thickness of the sample, and through an angle depending upon the ultimate strength; this angle to be not less than 90° at 28 tons, and 60° at 34 tons per square inch, and in proportion for strengths between those limits. At the present time all locomotive castings are made to the above specification when nothing otherwise has been specified, and all the engine castings for the Admiralty are made to the same limits. The castings to be thoroughly annealed for reasons already stated.

The wheel centres are required by some users to be rough turned, leaving ⅛ inch for finishing where they have to be finished bright. One wheel in fifty to be supplied by the

TABLE IV.

Test mark.	Length for torsion.	Diameter in inches.	Original area in sq. in.	Elastic limit in tons per sq. in.	Maximum stress in tons per sq. in.	Angle through which piece twisted for two diameters long.
						Deg.
S 74[1]	2·256	1·128	1·0	7·617	31·015	274
S 74[2]	„	„	„	8·161	30·470	275
S 74[1]	„	„	„	9·794	32·647	282
S 74[2]	„	„	„	9·794	29·926	213
013	„	„	„	7·610	30·400	278
014	„	„	„	9·790	31·010	271
015	„	„	„	7·610	31·550	285
016	„	„	„	8·700	32·640	316

TABLE V.

Test mark.	Authority.	Maker.	Length of Specimen.	Section.		Stress.		Strain.		Fracture.			Bends.	Remarks.
				Dimensions. Inches.	Area. Square inches.	Elastic limit. Tons per sq. in.	Breaking weight. Tons per sq. in.	Ultimate sets. Inches.	Per cent. of original length.	Area. Square inches.	Contraction per cent.	Appearance.	deg.	
C 337	Hadfield	Hadfield's Steel Foundry Co.	2	·7979	·5	—	32·5	—	29·0	—	40·4			
C 2¼ B			"	"	"	—	32·5	—	27·9	—	33·4			
C 147			"	"	"	—	34·5	—	27·7	—	35·2			
C 2¼ A			"	"	"	—	32·0	—	26·8	—	30·9			
C 111			"	"	"	—	40·0	—	19·0	—	20·0			
C 117			"	"	"	—	45·0	—	15·0	—	18·0			
C 1			"	"	"	—	52·0	—	4·0	—	6·3			
C 2			"	"	"	—	56·8	—	4·0	—	5·5			
C 3			"	"	"	—	64·0	—	6·0	—	8·0			
			"	—	—	—	26·0	—	36·5	—	40·4			1st centre arm } Loco. wheel centres.
			"	—	—	—	26·0	—	37·6	—	50·0			" " rim
			"	—	—	—	27·0	—	28·5	—	33·6			" " bal. weight
			"	—	—	—	26·5	—	20·6	—	22·3			2nd " arm

1958	Aspinall	Various	—	·7979	·5	—	36·9	—	—	·3632	27·3	Granular	35 to 80	
3221			2	·620	·301	—	30·0	·64	32·0	·1520	49·5	Fibrous		
3222			,,	·565	·250	—	32·5	·55	27·5	·1450	42·0	,, ...		
3223			,,	·540	·229	—	31·1	·60	30·0	·1190	48·0	,,		
3224			,,	·500	·196	—	32·6	·55	27·5	·1070	45·4	,,		
3435			,,	·560	·246	—	32·5	·27	13·5	·2200	10·5	Granular & fibrous		
3436			,,	·565	·250	—	31·2	·34	17·0	·2000	20·0	,, 30% ,,		
3437			,,	·565	·250	—	31·6	·60	30·0	·1350	46·0	Fibrous		
3438	Snowdon	John Rogerson & Co., Wolsingham, Darlington.	2	—	·237	18·6	32·4	·64	32·0	·1194	49·6	100% fibrous ..	180	Not broken. Size ½″ and [⅜″ radius
3439			,,	—	·216	18·5	32·9	·60	30·0	·1244	42·3	,, ,, ..	,,	Do.
3441			,,	—	·223	18·8	32·2	·62	31·0	·1152	48·3	,, ,, ..	,,	Do.
3442			,,	—	·227	17·3	31·7	·65	32·5	·1134	50·0	,, ,, ..	,,	Do.
3443			,,	—	·227	19·8	37·0	·45	22·5	·1661	26·8	20% fib., 80% gran.	,,	Do.
3444			,,	—	·225	18·3	31·6	·62	31·0	·1225	45·5	Fibrous	,,	Do.
3445			,,	—	·214	17·7	32·2	·60	30·0	·1134	46·9	,, ...	,,	Do.
3472			,,	—	·223	20·1	35·2	·50	25·0	·1520	31·8	4% fib., 96% gran.	115	Broken. Size ⅝″ and $\frac{7}{16}$″ [radius
3473			,,	—	·220	16·5	30·4	·65	32·5	·1134	48·0	100% fibrous ..	180	Not broken.
3474			,,	—	·212	19·6	32·4	·60	30·0	·1134	46·5	,, ,, ..	85	Broken.
3475			,,	—	·220	17·8	32·3	·65	32·5	·1194	45·7	,, ,, ..	180	Not broken.
3476			,,	—	·220	18·5	34·6	·52	26·0	·1520	30·9	10% fib., 90% gran.	100	Broken.
3477			,,	—	·224	18·4	32·0	·60	30·0	·1452	35·1	100% fibrous ..	180	Not broken.
3478			,,	—	·233	18·5	35·0	·42	21·0	·1960	15·8	100% granular ..	180	Do.
3479			,,	—	·220	18·6	31 4	·60	30·0	·1194	45·7	100% fibrous ..	180	Do.

TABLE VI.

Test mark.	Length of Specimen.	Section.		Stress.		Strain.		Fracture.			Bends without Fracture.	Remarks.	Passed by Inspector.
		Diameter. Inches.	Area. Square inches.	Elastic limit. Tons per square inch.	Ultimate load. Tons per square inch.	Extension. Inches.	Per cent.	Area. Square inches.	Contraction per cent.	Appearance.			
	in.										deg.		
Y 272	2	·76	·453	20·5	34·9	·65	32·5	·237	47·6	Silky	180	Engine castings	Admiralty
Y 282	"	·75	·441	20·6	34·7	·65	32·5	·212	51·9	"	180	"	"
Z 683	"	·77	·465	19·7	35·2	·60	30·0	·220	52·7	"	180	"	"
Z 689	"	·77	·465	19·3	33·3	·65	32·5	·246	47·5	"	180	"	"
Z 696	"	·78	·477	19·4	32·7	·70	35·0	·220	53·8	"	180	"	"
Z 703	"	·75	·441	19·9	35·3	·75	37·5	·196	55·5	"	180	"	"
Z 716	"	·80	·502	19·1	31·4	·75	37·5	·255	49·2	"	180	"	"
Z 719	"	·79	·490	18·1	31·8	·70	35·0	·229	53·2	"	180	"	"
Z 739	"	·76	·453	17·0	29·3	·70	35·0	·237	47·6	"	180	"	"
No. 1	"	1·13	1·002	19·8	33·8	·75	37·5	·502	49·8	"	180	"	Russian Government
" 7	"	·89	·622	18·0	30·4	·75	37·5	·301	51·6	"	110	"	" "
" 13	"	·88	·608	18·9	32·5	·70	35·0	·311	48·8	"	99	"	" "
" 15	"	1·11	·967	18·6	32·4	·72	36·0	·477	50·6	"	99	"	" "
" 10	"	·99	·769	18·3	32·5	·75	37·5	·390	50·7	"	100	Locomotive wheel	Railway
" 11	"	·99	·769	18·3	31·4	·70	35·0	·385	50·0	"	100	centres	"
" 12	"	1·00	·785	17·9	31·5	·70	35·0	·410	52·2	"	100	"	"
" 330	8	·88	·608	17·0	29·4	2·20	27·5	·311	48·8	"	90	Ship castings	Lloyds
" 353	"	·85	·567	17·4	29·2	2·00	25·0	·284	50·0	"	94	"	"
" 373	"	·85	·567	18·4	30·3	1·84	23·0	·295	52·0	"	127	"	"
" 374	"	·84	·554	18·1	29·9	2·40	30 0	·290	52·5	"	127	"	"

contractor free of cost, selected from the bulk and tested to destruction. From this sample the tensile and bending test will be taken, the results of which will be accepted as the average quality of the whole. It is further required by some users of wheel centres, that each be dropped in a running position, on the end of a spoke, from a height of 2 feet, and allowed to fall upon a rail secured to the top of an ingot of at least 2 tons weight, after which the wheel must be turned round through an angle of 90°, and dropped again in a similar manner. Each casting must be then slung up and hammered with a 7-lb. hammer to ascertain that it is sound. Clauses as to inspection and warranty are inserted to suit the purchaser; the latter is generally for twelve months from the date of commencing work.

As far as the analysis is concerned, probably it is best left in the hands of the steel maker, because so much depends upon the process. Sometimes a good analysis gives worse physical results than an inferior one, owing to the working of the charge during the earlier or latter stages of the heat; however, castings having an analysis of C ·18 to ·28, Si ·20 to ·40, S ·02 to ·07, P ·03 to ·07, Mn ·3 to ·75, would give similar results to those tabulated in Table V. under the second authority, providing that the heat had been worked to the best advantage. Finally, in a great number of observations made by the author for Mr. Aspinall, it was observed throughout, that the highest carbon with simultaneous high silicon gave the worst castings, and the best results were when the carbon was ·28 per cent., or below, or when the content of carbon was from 10 to 26 per cent. less than that of silicon. Perhaps this does not quite coincide with a comparison of Tables I. and II., but it must be remembered that these tables are given for specific reasons. The first, to show the effect of annealing and also chemical composition, and therefore various percentages of elements

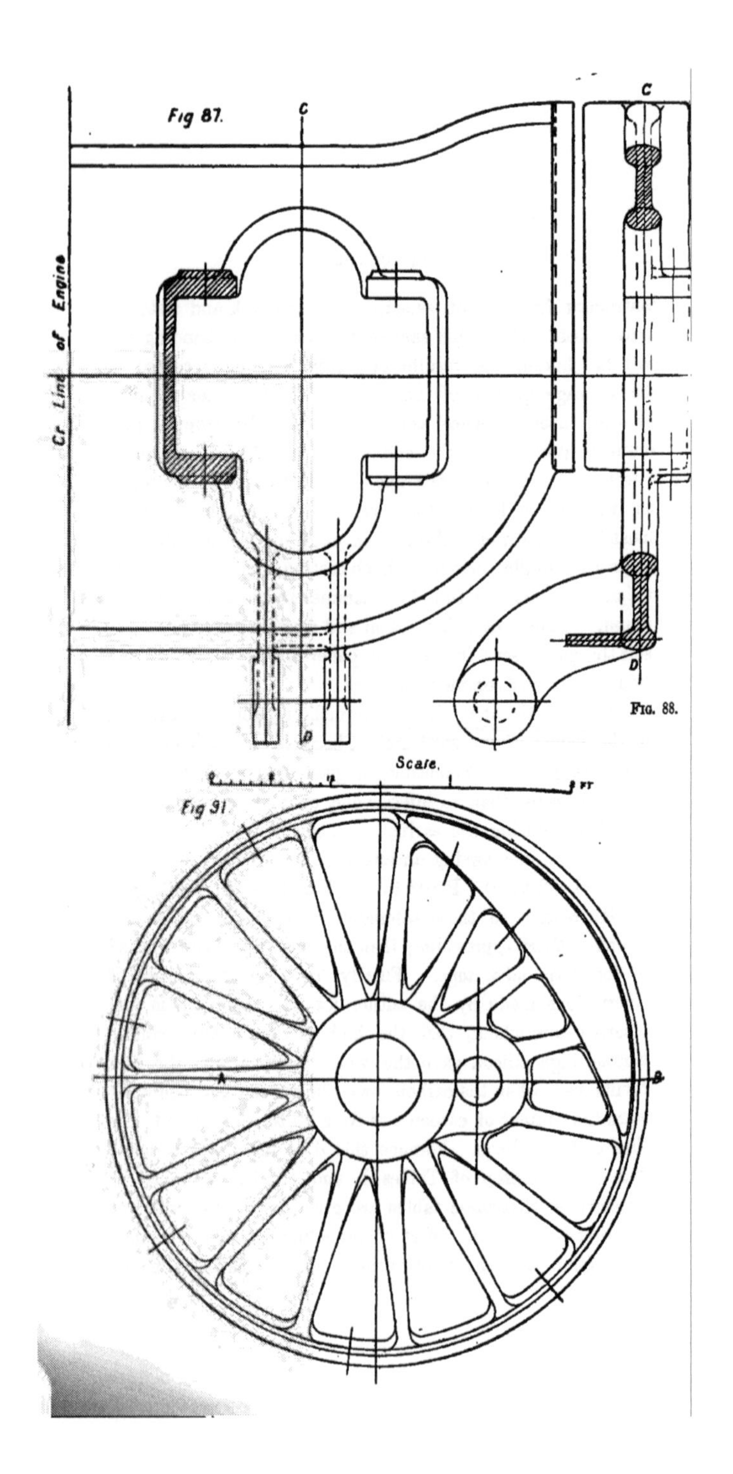
Fig 87.
C
Cr Line of Engine
D
C
D
Fig. 88.
Scale.
Fig 91
A
B

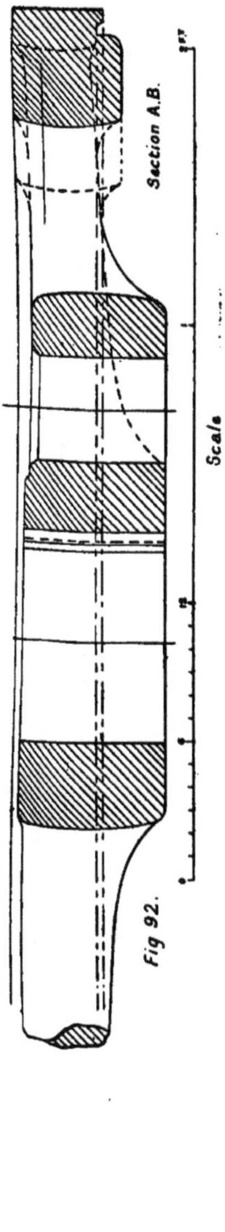

Fig 92.

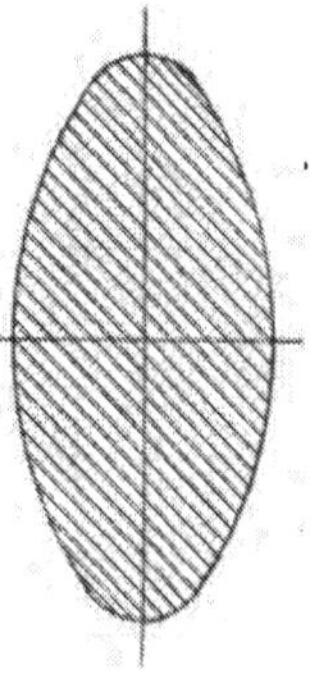

Section A.B.

Scale

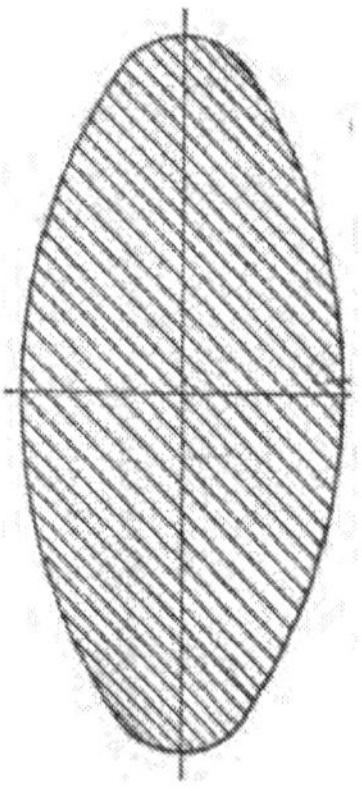

Fig. 93.

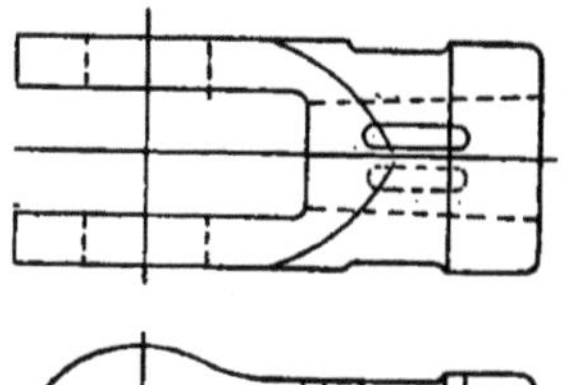

Fig 94.

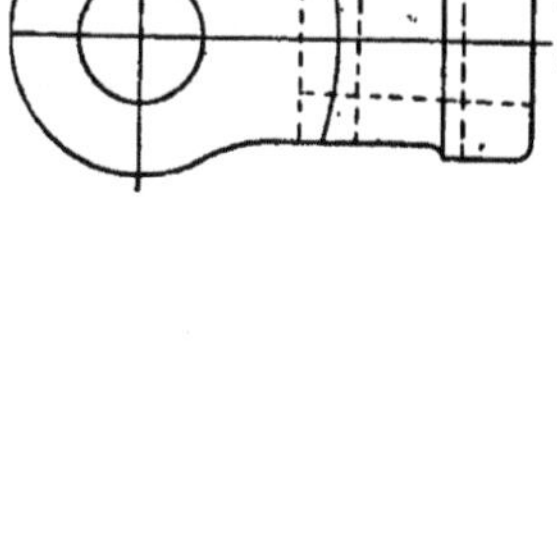

Fig 89

Fig 90.

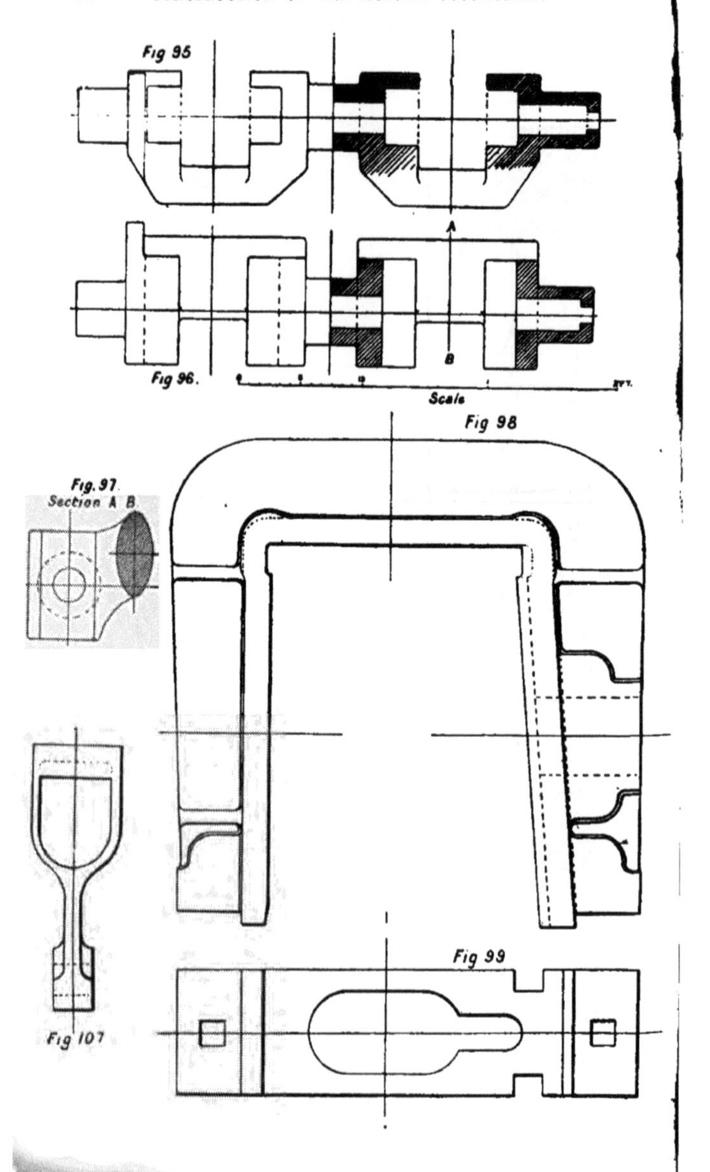
Fig 95
A
B
Fig 96.
Scale
Fig 98
Fig. 97.
Section A B.
Fig 99
Fig 107

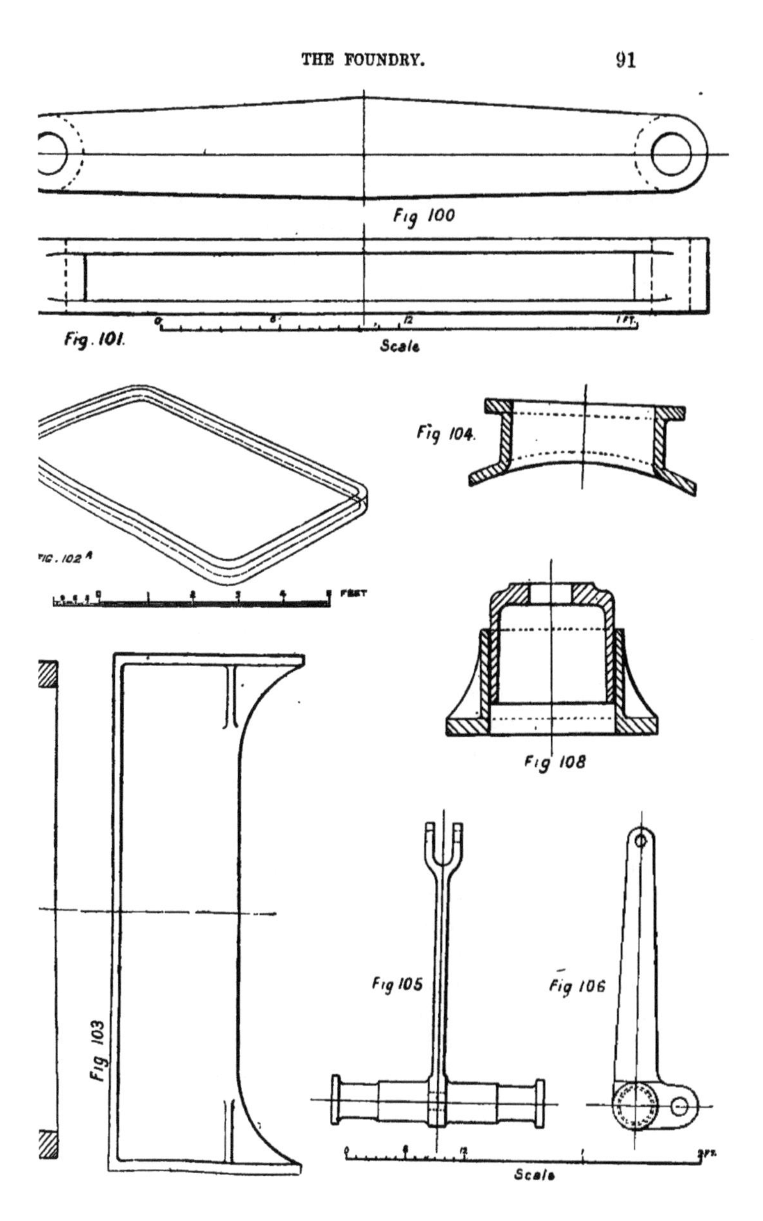
Fig 100
Fig. 101.
Scale
Fig 104.
FIG. 102 A
FEET
Fig 108
Fig 103
Fig 105
Fig 106
Scale

may be expected; whereas Table II. shows chiefly the remarkable reduction of the content of combined carbon by annealing. The reliability of these tables is unimpeachable, they satisfy a specific purpose, but the author's personal experience of the physical and analytical results of some hundreds of castings, is summed up by low carbon in conjunction with high silicon or *vice versâ*, these conditions being imperative, ·26 per cent. carbon being the highest preferable amount for mild castings. High silicon with low carbon will work admirably in the forge, with as much as 1·5 per cent. of the former, providing the latter is very low, say about ·15 per cent.

In Fig. 87 is seen one-half of the motion plate, from the centre line of the engine to the right frame, with its forked attachment for securing one end of the anchor link of Joy's motion, whilst an end view is shown in Fig. 88. Two views of the cross-head are given in Figs. 89 and 90, to which may be cast sufficient metal to be drawn down in the forge to the required dimensions of the piston rod. The driving wheel centre is given in Fig. 91, the leading and trailing being identical, with the exception of the balance weights, which are of course very much lighter; Fig. 92 being a half section through the crank pin boss. Figs. 93 and 94 are sections of the arms at the top and bottom. Reference has already been made to the designs of wheel centres, as well as the horn-block, which is shown in Fig. 98, with its keep, Fig. 99. This horn-block is of a useful design, but is frequently modified to suit the various types of locomotives built by different engineers. The Joy reversing shaft is shown in plan and elevation in Figs. 95, 96, and a cross-section, Fig. 97. Figs. 100 and 101 give two views of the brake hanger, which is swung from the frame at one end, coupled up to the brake cylinder by means of suitable connecting rods from the other, and having the cast-iron brake block attached in the centre,

Figs. 105 and 106 being the tender brake shaft. The fire-box foundation ring is given in cross-section in Fig. 102 and in isometric in Fig. 102A. The cross stay between the frames in front of the fire-box is shown in Fig. 103, and the safety valve seating in Fig. 104. Similar designs to Fig. 104 are adopted for the mud collector in the barrel, just in front of the fire-box, and also where cast-steel domes are used they require a similar seating. Fig. 107 is the tender spring link, and the draw bar spring box is shown in section in Fig. 108.

PART III.—BRASS FOUNDRY.

Copper and its alloys.—Copper has the capability of forming a combination with nearly all the metallic elements, producing compounds different in character from their component parts, possessing all the physical and chemical characteristics of metals, but often modified so as not to resemble either of those parts. This combination will either be of a complete chemical nature, or a solution of the chemical combination in an excess of one or the other. In itself copper possesses malleability and ductility to a considerable extent, which is clearly shown by the curve A, Fig. 108A. Its fluidity and tenacity is increased by a small addition of phosphorus, but unless alloyed with some other metal, such as tin, this addition of phosphorus increases its tendency to corrosion. The change in its physical properties, caused by the addition of zinc, is well illustrated by the curve B, Fig. 108A, each being autographic records of ordinary tensile tests taken by Aspinall's recorder. They do not require any further explanation, but those marked cohesive force have been plotted from the curve of tensile strain, utilising the data given by Professor Unwin on page 419, vol. lix. of the *Engineer*, where he shows that assuming the sample to be uniformly plastic throughout its length, which for our purpose is approximately correct up to its climax of resistance, the percentage of contraction of area is equal to the percentage elongation, calculated upon the stretched

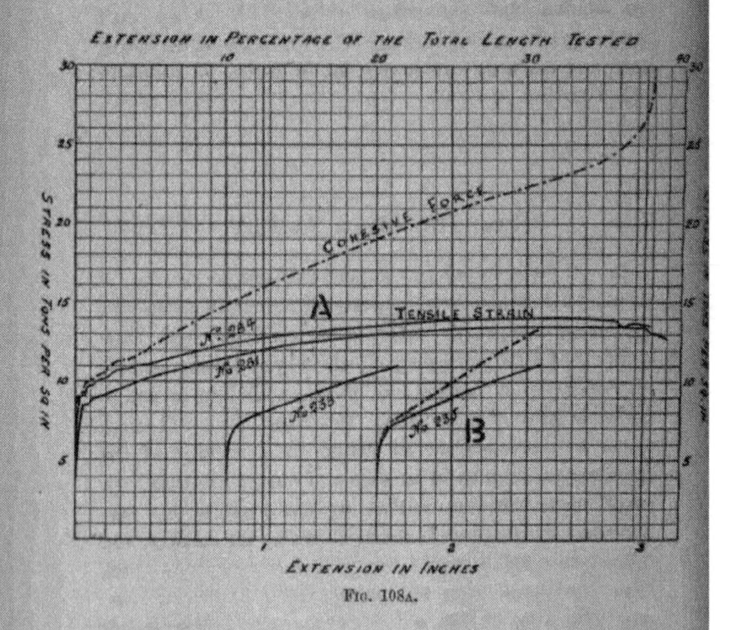

FIG. 108A.

length of the bar. In the examples, this point has been ascertained, a new scale constructed, and the curve has been plotted in a convenient manner by raising the climax of resistance by the required percentage. Oblique lines have then been ruled across to the original scale, and the curve plotted by noting where the load curve crosses the load ordinates, and transferring this point vertically to the corresponding oblique line. The remainder of the curve is obtained by calipering the contracted area of the sample, which fixes the highest point; the intermediate ones however cannot be obtained, because the length subject to local contraction becomes more and more restricted as the moment of the fracture approaches, therefore the path of this portion of the curve is merely hypothetical. A synopsis of the particulars of these curves is given in Table I. Annealing the copper samples before testing would have produced 45 per cent. elongation.

TABLE I.

Curve.	Test No.	Stress in tons per sq. in.	Elongation per cent. on 8 inches.	Contraction per cent.
A	281	13·66	40·0	52·9
"	284	14·04	38·5	53·8
B	233	11·00	11·5	20·9
"	235	11·00	11·0	20·9

In the manufacture of homogeneous alloys, the following may be taken as characteristic difficulties, with their total or partial solution. Impurities in the diluting metals, which will seriously injure the physical properties of the resulting alloy, decreasing as the purity of the diluting metals increases by improved metallurgical methods of manufacture. The oxidation of the metals to an always varying extent, which occurs during the period of melting, and will always remain a difficulty, only to be mitigated by experience of the furnace, fuel, and quick melting. The occlusion of oxygen, the combination or mixture of oxides, produced from the constituting metals during melting, with the alloy, or the absorption of small quantities of the products of combustion, including carbon and sulphur, and then the evolution of these gases or expulsion of oxides during cooling, rendering the alloy porous, which may be remedied by the addition of phosphorus. Finally, liquation at the period of cooling, when the alloy will separate out in distinct groups of various grades of hardness, unequally diffused throughout the mass of the casting, which can be remedied to a certain extent by adopting means for rapid cooling. The larger portion of the heavier metals will assume a level in the crucible in order of their respective densities, unless they are prevented from doing so by violent agitation, and a partial separation will consequently take place in the casting, even when all the attention possible may have been bestowed upon the alloy during its formation, especially if the casting is large, and therefore the cooling slow, also if large proportions of lead or tin have been used; therefore, to obtain perfect homogeneous alloys, remelting is sometimes a great advantage, especially if it can be accomplished with the minimum amount of oxidation and occlusion going on. If there has been an appreciable amount of oxidation it is impossible to get good alloys, because, if the author may illustrate by Dalton's atomic theory, their atomic bonds are clogged with this slag, and consequently the metals will not "wet" each other.

Bronzes, copper-tin alloy.—Copper and tin mix well in almost all proportions, forming a class of alloys generally spoken of as bronzes, sometimes having incorporated with them lead, antimony, manganese and phosphorus, with iron and silicon as impurities. A small content of tin renders the alloy both hard and tenacious, a maximum hardness for all shop purposes being attained by the addition of about 15 to 18 per cent. Table II. fairly represents the various mixtures of these alloys.

These alloys have a lower melting point than copper, a greater density than the mean density of the chief constituents, less liable to oxidation, and they are proportionately harder than either of the principal constituting metals, especially as the content of tin increases. The fact of their being more fusible than copper and less than tin, renders it very difficult to obtain a perfectly homogeneous alloy, portions richer in tin being interposed through the mass; therefore it should be cooled as rapidly as possible, to obviate any tendency to liquation during the cooling. Oxidation must be avoided as far as possible, which points to the rapid melting of the copper and keeping the tin immersed during

TABLE II.

Description.	Cu per cent.	Sn per cent.	Zn per cent.	Pb per cent.	P per cent.	Fe per cent.	Si per cent.
Old copper-tin alloy	87·5	12·5					
A standard American bearing	80·0	10·0		9·5	·5		
English generally	80 to 88·0	18 to 10	2 2	·5 ·5			
Particular cases of slide valves	82·5	14·5			·36	2·0	·04
	85·0	11·6	3·9		·16		
	83·5	6·0		10	2 to ·4		

mixing, because of the rapid formation of the peroxide. This formation is very disadvantageous, becoming a great nuisance to the finisher, rendering all machine work imperfect, because of the hard dirty spots, which destroy the cutting edge of the tools. Especially is this the case if the metal has been produced by the open hearth Siemens regenerative furnace, when that furnace has been working slowly; the metal will be found to contain a greater quantity of these hard dirty spots, also there will be more wasters in castings.

The small quantity of zinc used, say 2 per cent., plays the beneficial part of deoxidant, by reducing the incorporated oxides, rendering the product purer, consequently of greater tenacity, but at the same time it has a tendency to lower the elastic limit and soften the alloy. Phosphorus behaves in a similar manner to zinc, but with greater energy, and in increased proportions it sensibly hardens the alloy. Phosphor bronze is generally manufactured either by the addition of copper or tin phosphide, and after having eliminated the oxides, a further increment to produce from ·25 per cent. to 2·5 per cent. in the finished product will change the colour to a greater degree of evenness, the fracture will be finer, and its physical properties and fluidity at the time of casting will be greatly increased. In fact, the use of phosphorus in the brass foundry may be compared to carbon in the steel foundry.

The chief uses of phosphor bronze may be taken as slide valve and bearing metals, the former a most important item on a railroad. To obtain a standard slide valve mixture, valves of an experimental mixing should be made with varying contents of phosphorus, noting at the time of manufacture all the points connected therewith, as much depends upon the facilities of the shop and the skill brought to bear upon their production. Some of the points worth noticing are the condition of the furnace, length of time in melting the copper, and approximate temperature at the time of casting. The fracture of a small ingot should be reserved, and where possible mechanical and analytical tests should be made. The behaviour of these valves should be watched and compared with the wear of the ordinary mixture. In this class of work good clean copper shearings and new metals are resorted to, but for general work the accumulation of old valves has to be dealt with, or it may be that all the old scrap is melted in an open-hearth furnace and cast into ingots. In any case the scrap has to be disposed of, and in only one obvious manner. The quantity of scrap used in the mixings will of course regulate itself to the quantity on stock, say up to 50 per cent. of the charge. The use of this scrap is a very easy matter if its nature is thoroughly known, and consequently reduces the quantity of new metal required; but if this is not known, then the mixing must be based upon, say, an average analysis of half-a-dozen samples of the scrap. The addition of the phosphide must be the last operation, after lead and tin, the copper being melted as rapidly as possible for reasons already stated, that is, to lessen the work of the phosphides, the whole being afterwards well stirred and poured. The knowledge necessary for pouring at the right moment is very soon acquired by the operator, by watching the working of the metal and the rising of the scum, the latter having to be removed. In the case of bearings, for durability alone, they should be as hard as the axles they support; but considering the wear of the latter, the former should be softer, so that the wear, say, per one thousand miles is about three to one. Owing to the difficulty in obtaining homogeneous alloys, it often happens that in liquation the harder alloys separate out, and form the interior metal of the bearing instead of the outside casing, resulting in having for the actual bearing a soft alloy which rapidly wears, then the axle coming upon the hard places, causes, in the absence of ample lubrication, its destruction. The soft alloy cools first and forms the shell, the harder filling the interior spaces, which will probably contain double the content of tin to the former. Everything points to the use of a rich alloy of phosphide just before the time of casting, well stirring, and rapid cooling. A fluid pressure caused by a large gate and head, will enhance its resistance to compression. It has also been observed that those alloys which wear the longest have a low tenacity and a good elongation; but by retaining the elongation with an increased tenacity, the metal would have double reasons for increased durability. The granular structure is also an important factor in the wear of metals, because the finer the structure the finer will be the flaky abrasion, and consequently longer wear. Among the many conditions which will affect the wear may be enumerated lubrication, resistance to abrasion, pressure, speed, and in the case of slide valves, temperature; but supposing these conditions to remain constant, then the resistance to wear must be sought for and applied in the alloy itself, and it will always be found that with any ordinary shop mixture, if precaution be taken to produce it with an increased amount of purity, it will resist wear to a greater extent, this probably being done by the small addition of zinc or phosphorus to remove the oxides, and beyond, by the addition of a greater amount of phosphorus, its wearing properties will be enhanced. It has also been found that the increased quantity of lead, diminishing tin in proportion, has a very beneficial effect upon the wear of this class of alloys, and as a matter of fact, 10 per cent. of lead and only 6 per cent. of tin is regularly used for the slide valves of the engines in question, with about ·25 per cent. to ·3 per cent. phosphorus.

Brass.—The alloys of copper and zinc may generally be accepted under the term of brass. They are made in great variety of physical properties and appearances, varying from the colour of the latter metal to that of gold. The content of copper may be from 60 to 90 per cent., but generally the composition of English brass is about 70 per cent. copper and 30 per cent. zinc, with about ·5 per cent. lead, which will make it work better in the machine shop, and is generally found in locomotive brass. The best proportion for yellow brass is two of copper and one of zinc. When mixing, an allowance of about 2 per cent. must always be made for the elimination of zinc, caused by its point of volatilisation being very low, and for a similar reason—its low melting point—brass cannot be forged. Zinc produces the hardness, which gives the alloy a greater resistance to wear, being at the same time malleable and ductile. If slowly cooled it is more hard and brittle than if plunged whilst at a red heat into cold water.

White metals.—White and antifriction metals are almost without number, and until recently their base was tin—in some cases to the extent of 78 to 86 per cent. Now, generally speaking, lead has taken the place of tin as a base, with from 10 to 20 per cent. of antimony as a hardener, and sometimes arsenic. Copper is in most cases added to the extent of from 2 to 10 per cent. It will be found, as in the case of brass and bronze, that every shop has its own mixing, which is quite as good as any floated upon the market, and mostly having the same ingredients, and nearly equal percentages. The chief object in a good metal is to have a low melting point, especially where there is a lot of work, so that it can be used over a pot fire, that is, without the aid of crucibles. It should be thin when melted, but slow to set, having a lengthy plastic stage, so that the workman can follow it up in a similar manner to wiping a plumbing joint when soldering two lead pipes.

Other Alloys.—After these two distinct alloys of copper, brass and bronze, the rest may be disposed of as entirely of an experimental nature, although many, such as ferro up to 3·5 per cent., silicon, manganese, and aluminium bronzes, are being used extensively. In the case of manganese bronze, ferro-manganese is added to the molten copper, the principal use of the manganese being to free the copper from incorporated oxides, which rise to the surface as slag. If more manganese is used than is sufficient to remove the oxides, the alloy is of greater tenacity and more ductile, the maximum being attained with 10 per cent., beyond this amount the alloy becomes very hard and brittle, until at 15 per cent. it is of a grey colour. Aluminium bronze may be obtained by using good copper and re-melting; 90 per cent. copper and 10 per cent. aluminium having a definite chemical composition, Cu_8 Al. There is a difficulty in getting the aluminium incorporated with the copper, also in pouring, as the alloy sets very rapidly. The best method, as found by many experiments carried out by the author for Mr. Aspinall, is to make the above rich alloy, cast into ingots, and use in an ordinary manner with the usual mixture. A 3 per cent. aluminium bronze gives very good results—double the tenacity of good brass—and at very low temperatures it will forge well, aluminium itself becoming red-short at 400° F. This 3 per cent. bronze would be very useful for castings, although it cannot compete with steel as far as first cost of material is concerned; but it has this advantage, that only working parts need be machined, as it makes a very smooth surface. Aluminium passes through a granular stage, and becomes fluid at about 1300° F., copper melting at 1980° F. It is advisable to use the best plumbago crucibles, because from silicious ones it becomes seriously contaminated with silicon. A flux is not necessary, and might become detrimental, owing to the extreme lightness of the metal. At the time of casting a skimmer should be used, in order to remove the thin film of oxide which always forms; and then it should be poured quickly. Larger gates are required, and the best castings are obtained from dry sand moulds, although very good ones may be obtained by skin drying only. Coruscation is characteristic of the introduction of aluminium into the crucible, and the castings sometimes present locally an iridescent appearance.

Melting.—This will either be crucible or open-hearth. A good crucible must be able to withstand the effects of the chill when drawn from a furnace of high temperature and immediately exposed to the atmosphere, after pouring. It must also be sufficiently refractory to withstand the highest temperatures, and it should not impart any of itself to an appreciable extent to the ingredients which are being melted in it. Plumbago crucibles are the most satisfactory in all respects, generally withstanding over forty charges before destruction. They can be used in a coke or gas furnace, the latter on the regenerative system. The former may be made either square or round, with wrought or cast-iron plates, lined with either fire brick or clay, about 18 inches internal diameter and 4 feet deep, with ordinary furnace bars, sunken ash-pit, high stack, and the top of the furnaces being on about the floor level. Although the regenerative crucible furnace for brass melting has not been generally adopted, objections being raised against the use of gas and oxidation, it has not been found that the difficulties are beyond surmounting. Experiments carried out by the author for Mr. Aspinall show a loss of 2·5 per cent. from oxidation. The furnaces are constructed to hold from fifteen to thirty 100 lb. pots, very similar to the arrangement adopted for crucible steel. They require overhauling about twice a year, when the chambers can be re-lined or chequers removed and a thorough repair accomplished in a day or two. A useful size for the open hearth is about 30 to 40 cwt. The charge would then consist of about 15 to 20 cwt. of brass scrap, including turnings, broken scrap, shot metal obtained from hot skimmings by grinding and washing, and about 5 cwt. of good copper scrap. It is worked by adding new metal, zinc, tin and copper, as the case may be, according to the nature of the fracture, which has been sampled from the molten metal. This should be close, homogeneous, fine grain and of even colour, zinc or copper being added until a satisfactory fracture has been obtained. This is a necessary mode of procedure, owing to the nature of the charge and the elimination of the zinc, resembling very much the sampling of a molten bath of steel.

Moulding.—In the present section, dealing with the moulder's work in the brass foundry, in order to entirely separate and distinguish it from that in the iron foundry, the illustrations are of the patterns used and not their imprint in the sand. They fairly represent both the pattern-shop and brass foundry practice, and in themselves are sufficiently clear to require but brief explanation. All core prints and

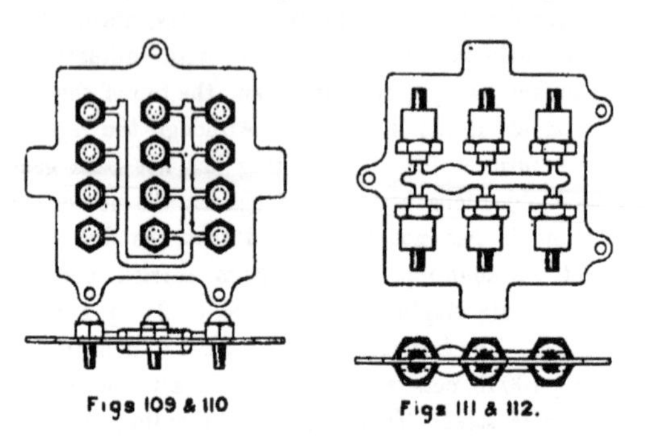

Figs 109 & 110 Figs 111 & 112.

bearings are indicated by an irregular diagonal shading, whereas the regular shading represents that portion of the pattern or core-box to be metallic. The range for plate work is even more infinite than in the iron foundry. All descriptions of brass nuts, such as required in the smoke-box, lead and mud plugs, are of this class of work, illustrated by Figs. 109 and 110, by 1 inch nuts, the sizes ranging to 2½ inches; each plate having from two or three to thirty articles upon it, according to the dimensions of the pattern. They are placed in such a position that the main body of the casting is in the top box, whereas the slide bar oil syphons and cylinder cocks are arranged as indicated by Figs. 111 to 114. In each of these cases the plate is placed upon a turn-over board, sprinkled with parting sand, rammed up, turned over, and the operation repeated, with due regard to venting. The runner or leader is formed by ramming in a suitable piece of wood which, after streaking over with a moist rag, is drawn. In Figs. 109–114 the gates are at once discerned, and should the furnace be working slowly, and consequently cold metal expected, they are made larger by suitable means, apparent to the reader, thereby giving the castings a better

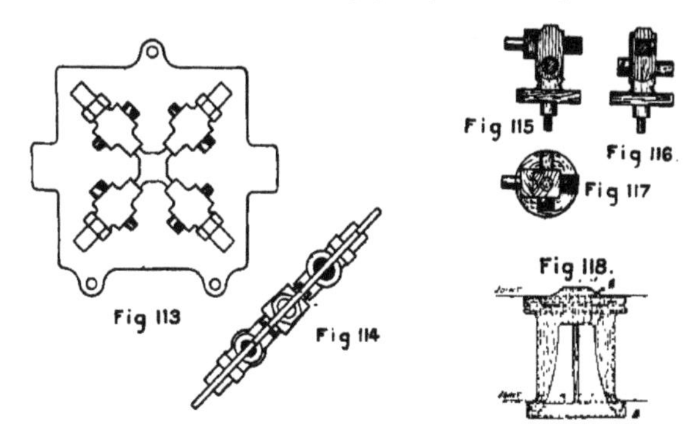

Fig 113 Fig 114 Fig 115 Fig 116 Fig 117 Fig 118.

feed. Figs. 115, 116, and 117 are the patterns for the water gauge cock, and although they are shown as made of wood, they are one of those articles that can be placed upon a plate and cast with the core print A in the bottom box. Fig. 118 shows the pattern of Ramsbottom's regulator valve, forming its own cores. For this pattern three boxes are necessary, the middle box receiving the upper portion of the pattern, minus the wings A, which is placed upon a turnover board, rammed, turned, and the bottom joint adjusted, then the bottom flange B is rammed, the whole turned, and the wing piece A pushed in and the top box rammed. When each box is removed, it will be clearly seen that the wing piece has made an impression in the top box, the body in the middle, and the bottom flange, or seating, in the bottom. It is cast by means of an inch gate on the top and bottom flanges. An alternative plan is to have a pattern lagged up, with suitable core prints and joint parallel with its

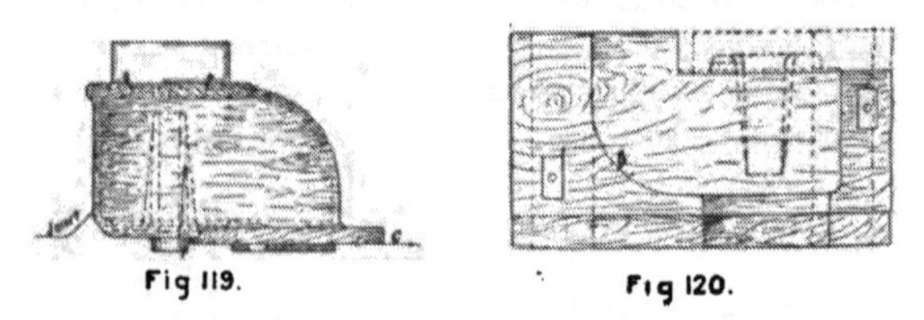

Fig 119. Fig 120.

axis. This would necessitate a separate core-box; but for many reasons the former is the better example, chiefly from the fact that it makes its own core, which consequently cannot be misplaced and otherwise distorted when being fixed by an indifferent moulder.

The pattern for the seating of this valve is shown in

Fig 121. Fig 122 Fig 123

Fig. 119, and its core-boxes are shown in Figs. 120–123. The pattern is moulded in a suitable box placed on a turn-over board suitably adapted, A, the joint being made when turned over, as indicated by Fig. 119. Fig. 120 is an elevation of the body core on its joint, with Figs. 122 and 123 shown in position in dotted lines, which forms one of the seatings and the guide for the valve, Fig. 118, A, Fig. 123, being a strengthening piece through which one of the holding-down bolts passes. Fig. 121 is a plan of the body core-box, with Figs. 122 and 123 removed. One chaplet is used under the unsupported end of the core, as indicated by B, Fig. 120. It lies sideways, and its sprig rests upon a plate which has been rammed up in the bottom box. The casting is run at the flange C by means of a rectangular gate, 1½ inch by 1¼ inch. Two views of the pattern for the regulator rod

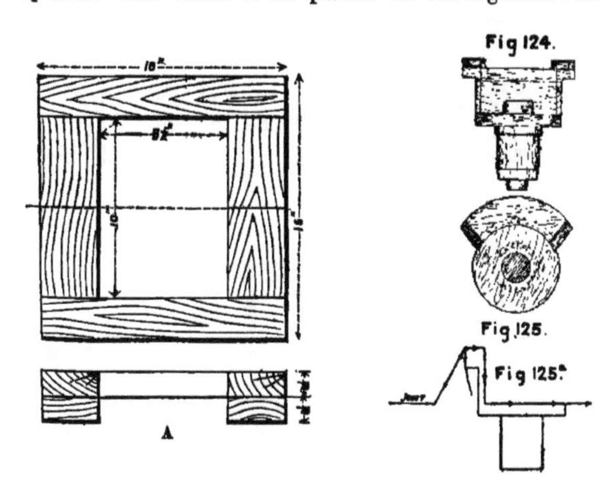

stuffing-box are given in Figs. 124 and 125, the joint being made along the flange, as indicated by the sketch, Fig. 125A. To facilitate the moulding of this pattern, as in several other cases, an "odd side" is used; that is, after one mould has been run, instead of knocking both boxes out, one is retained as a kind of print, to receive one portion of the pattern, until the opposite portion has been rammed up; or, in some cases, an odd side is specially made. The gate is attached to the flange. The gland and brass sleeve for this rod are shown in

Figs. 126–128, all glands being cast down, and when they form their own lightening cores, these are hardened by skin drying.

All sleeves, Fig. 128, and the bush for the blank gland in the steam chest front cover, Fig. 145, are moulded for a straight draw if only one core print, or if two core prints they are moulded horizontally. Fig. 129 is the cone for the joint of internal steam pipe. It is cast with a 1½ inch

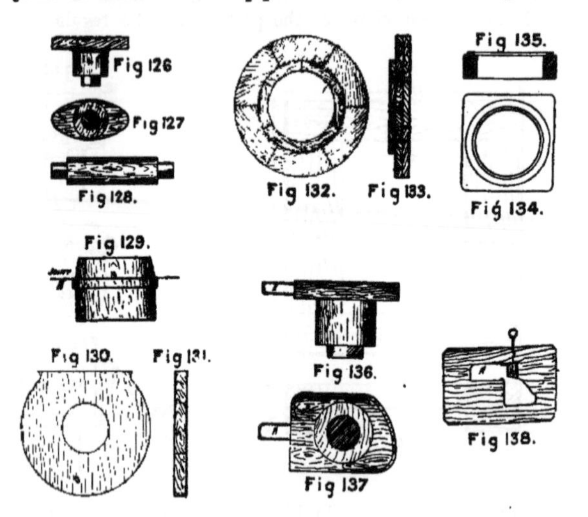

runner on to the edge at A, all the core being left in the bottom box and the joint, as shown in Fig. 129. An odd side is also used to expedite the moulding of this pattern, and it is fairly vented in the top box. Figs. 130–135 are various flanges for the steam pipes. They are all cast in the bottom box, with 2-inch runners on to their edges, and make their own cores. Figs. 136 and 137 is the piston-rod gland, and Fig. 138 its core-box for the oil syphon, A being the core prints. The gland for the valve-rod is shown in Figs. 139 and 140, and its core-box in Figs. 141 and 142. These patterns form their own lightening cores by being hollowed out, and rammed with sand, their method of moulding having already been indicated under the heading of regulator

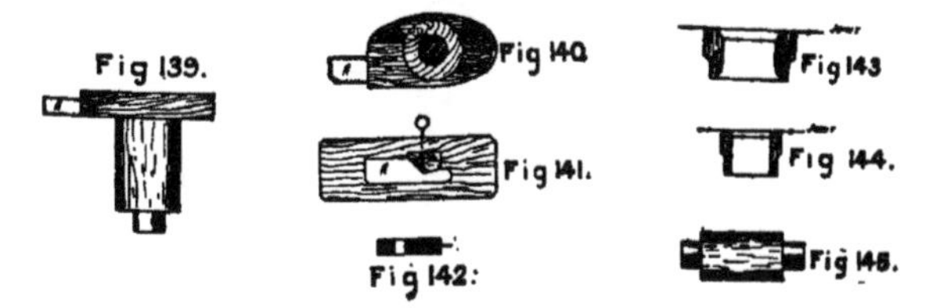

gland. Their bushes are given in Figs. 143 and 144, the joints being as shown.

Figs. 146–148 give views of the slide valve, the importance of which has already been demonstrated, especially relating to the metal from which it is cast. By adopting the following method of moulding the slide valves, all necessity for machine work upon these castings, with the exception of grinding the faces over, is dispensed with, which reduces

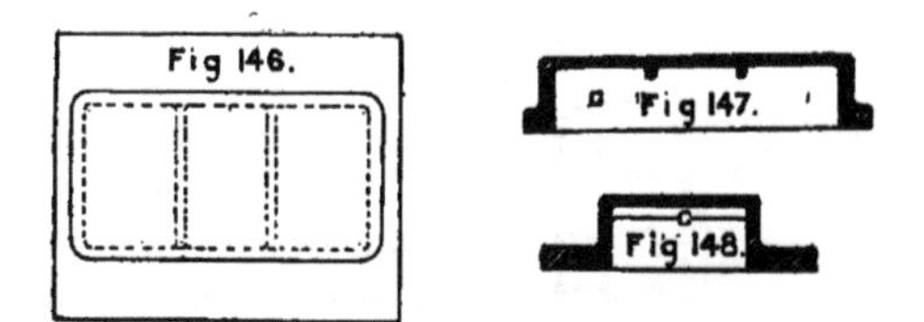

their first cost considerably. The exact thing is not arrived at all at once, but after a few trials and a knowledge of the behaviour of the mixture, the patterns can be so regulated that the shrinkage will be sufficient for the resulting casting to fit the standard gauges. If the mixing should be changed (which is very seldom when once a good working one has been obtained, economical in wear and other points), the patterns would have to be lined up, or dimensions reduced, according to the amount and direction of contraction. The pattern, which is metallic, is placed upon the

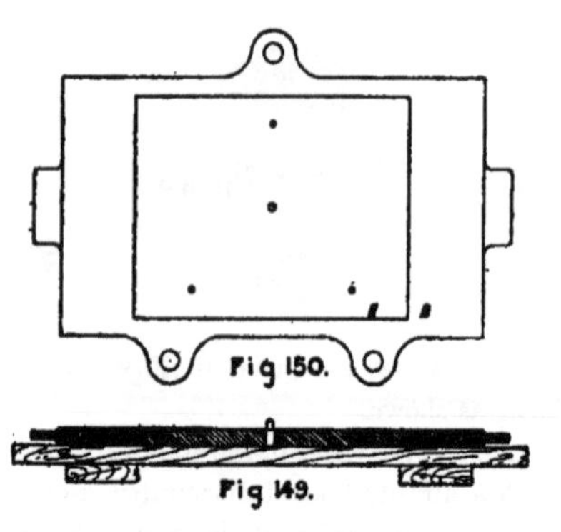

plate A, which is a good fit in the plate B, but slightly thinner, each plate having suitable dowels, Figs. 149 and 150. They are placed on a turnover board, the top and bottom boxes rammed up and vented, the print of the pattern being in the top box. The plate A being slightly thinner than the plate B, the flange of the pattern just enters, consequently when the bottom box is removed and the plate A drawn, the pattern can be drawn upwards through the plate B with a minimum amount of

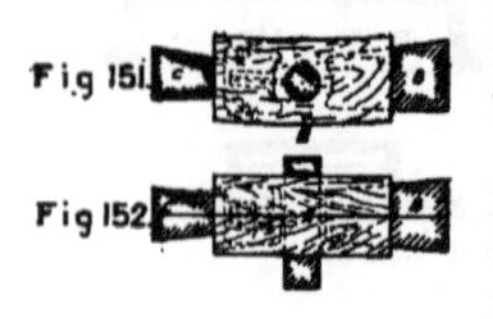

swelling to the mould. The core for C, Figs. 147 and 148, is formed by ramming up that space in the pattern and joining it to the bottom box, through the plate B, the casting being of phosphor bronze. The radius or slipper blocks for Joy's valve motion are also of phosphor bronze, the patterns of which are shown in Figs. 151 and 152, the half with the short core print being placed in the bottom box, well vented in the top and slightly in the bottom, and run at the joint at A. The lightening core B, Figs. 151 and 152, is given in Figs. 153 and 154, and the oil syphon C, Figs. 151 and 152, in Figs. 155 and 156.

When the axle-boxes are not cast in steel or iron, Figs. 157–162 show the patterns and core-boxes for the brass foundry. The core print A is placed upon a turnover board, and the joints made at B and C, Fig. 157. The metal strips,

Fig 153 Fig 154

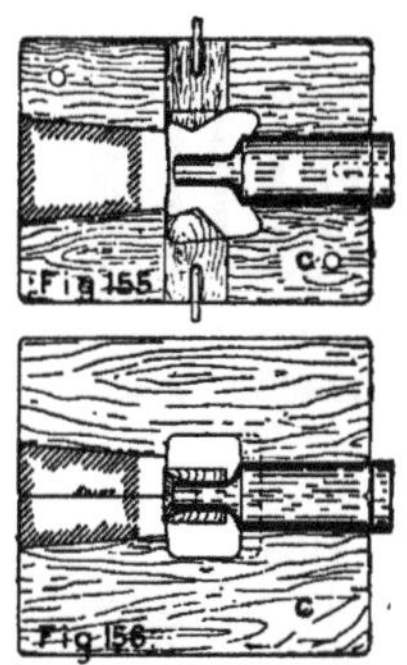

Fig 155 Fig 156

Fig. 159, are bevel pieces let into the sand to form the corners at the top of the box, where it enters the horn. Fig. 160 is the core-box for the oil syphon and tallow receiver, and is suspended by a wire from the outside of the top box to the core iron, and fitting into the core print A. Figs. 161 and 162 are two views of the core-box for the white metal recess, and correspond to the core print D. The casting is run by two branches from each of two leaders at the top and bottom of the mould, the leaders being rectangular, $1\frac{1}{4}$ inch by $1\frac{1}{8}$ inch, tapering to $\frac{7}{8}$ inch by $\frac{3}{4}$ inch. The tender axle-box brass is shown in Figs. 163–168, being top and bottom plans, side and end elevations. The metal strip A, Figs. 163 and 166, enables the pattern to form its own white

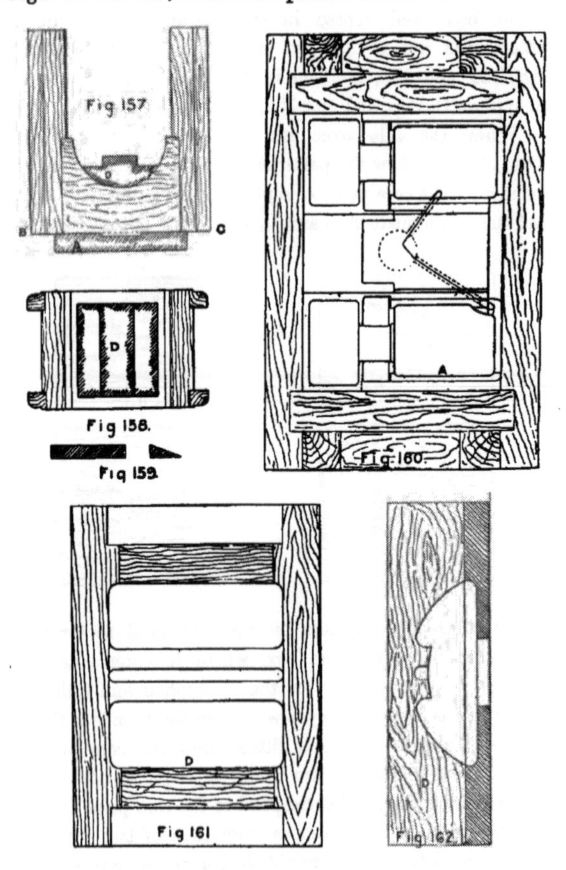

metal cores. The joint follows the core prints at the sides and the curve of the bearing between, in order to allow of a free draw. The casting is run on to the top of the mould at one end about the middle, see B, Fig. 165, this also indicating

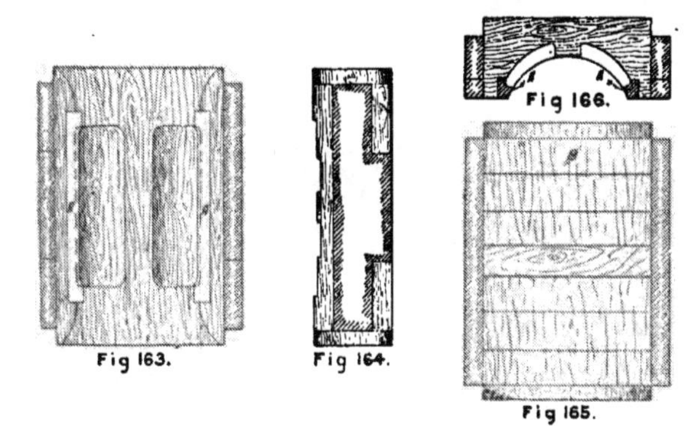
Fig 163. Fig 164. Fig 166. Fig 165.

the position of the pattern in the box, which is well vented in the top and slightly in the bottom. Fig. 169 is the pattern for the safety valve pillar, and Fig. 170 its core-box, the

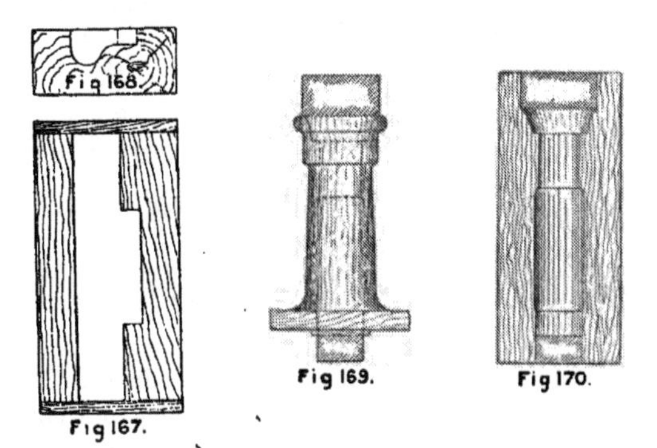
Fig 168. Fig 167. Fig 169. Fig 170.

joint being a diagonal with the bottom square. The scoop for Ramsbottom's water pick-up is shown in side elevation,

Fig. 171, the joint being indicated from A to B along the centre of the pattern. Fig. 172 is a plan, Fig. 173 an end elevation, and Fig. 174 its core-box. It is cast in boxes specially adapted for the work; the loose parts held by pegs

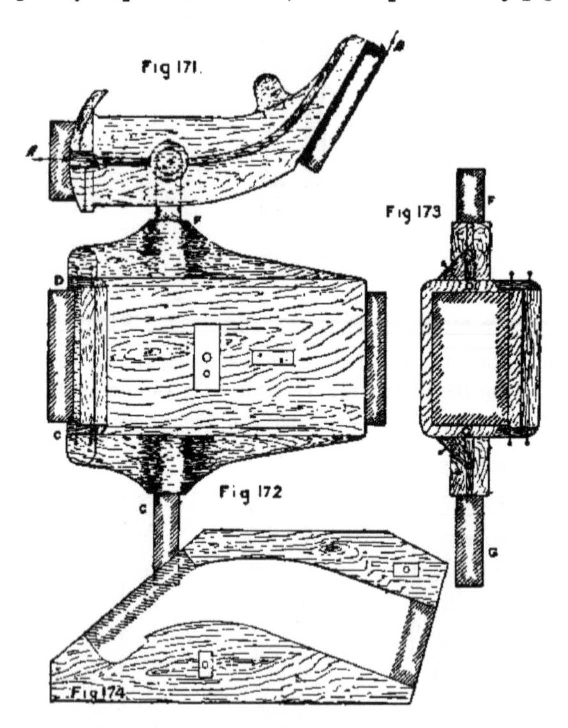

are rammed up slightly, and then the pegs are drawn. It is run from two leaders at D and C, F and G being core prints to form bearings for the wrought-iron arms which are cast in. Fig. 175 is a finished sectional drawing of the ball-and-socket connection between the tender and the injector, and

Figs. 176–186 its pattern in detail, which indicates clearly the method of procedure in moulding.

The No. 8 combination injector is given in Figs. 187 and 187A, its pattern in Figs. 210 and 211, the core-boxes in Figs. 212–217, and patterns of details in Figs. 188–209. It may be observed that the shop pattern number is 2710, and

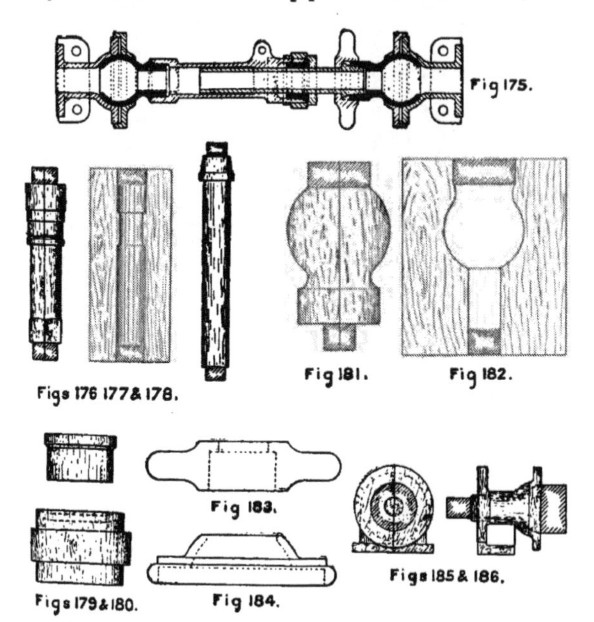

Fig 175.

Figs 176 177 & 178.

Fig 181.

Fig 182.

Fig 183.

Figs 179 & 180.

Fig 184.

Figs 185 & 186.

the details range from 2711–2730, so that comparing the drawing with the details, it will at once be seen how they are moulded. The core-boxes, Figs. 212–217, are lettered A to D, and the finished drawing has also the corresponding letters, so that each core can be identified with that portion of the mould.

It may be considered an opportune moment to describe this injector, as when dealing with it in the machine shop it

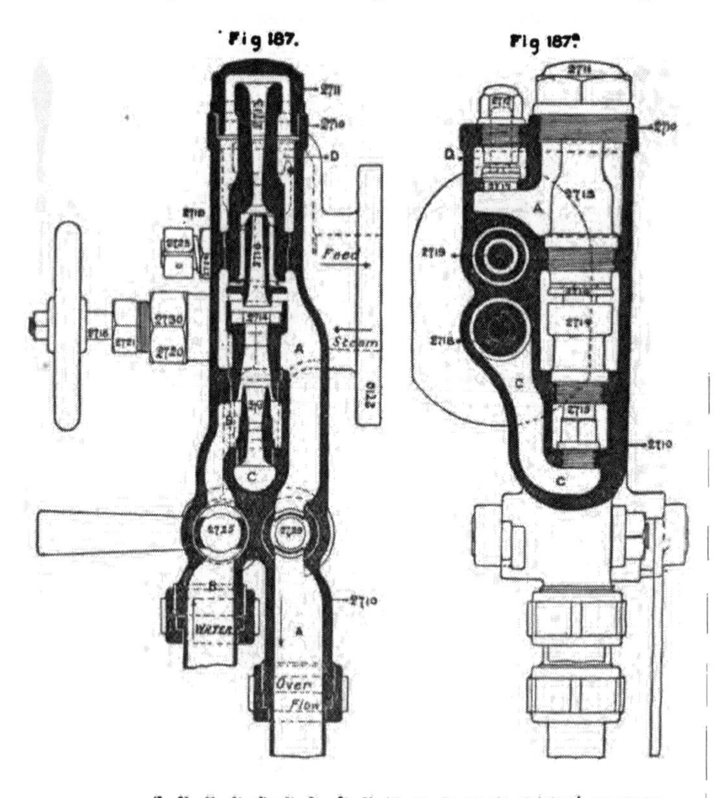

No 8 Combination Injector.

will then only be necessary to describe the tool work required. The patent rights claiming improved construction are held by Messrs. Gresham and Craven of Salford. The casing of the injector, along with the flange junction for the boiler attachment, is formed in one piece. The mechanism constituting an injector, viz. nozzles, cones, valves, cocks and other parts, is combined together so as to utilise this one flange or equivalent connection, having two openings, one for steam and one for water, and fixed to the fire-box front with one joint, the openings being each connected with a pipe leading to the points required in the interior of the boiler.

Fig. 187 is a sectional elevation on a plane at right angles with the face of the flange forming the junction with the boiler, the section being taken through the water and overflow cocks, and the axis of the nozzles of the injector. Fig. 187A is a sectional elevation on a plane parallel with the face of the flange, transverse to the axis of the steam and delivery cocks, and showing the nozzles of the injector in simple elevation. Within the flange, by means of which the injector is secured to the boiler, there are two passages, that marked "steam" being in communication with the steam space of the boiler, and that marked "feed" being connected with a pipe, the open end of which delivers the water in the required part of the boiler. The steam valve 2718 has a circular chamber 2720 or 2730, as the case may be, and stuffing gland 2721. This steam valve may either have a single seat, as shown, or double, the latter case being especially adapted for the use of steam for other purposes, such as the vacuum brake ejector, and steam sanding arrangements. When the spindle is screwed out to its full extent the steam has access to the casing of the injector only, that part enclosing the nozzles being shut off by the steam valve coming upon its back seating. The pattern 2722 is only required in the latter case, and is to prevent the spindle from being screwed so far as to bring the valve close against its seating, and thus shutting the steam from the boiler entirely off from the injector casing, including the cones, also giving latitude for regulating the steam within the limit fixed by the arm 2722. 2719 is the stop valve in its casing 2724, having a square nut

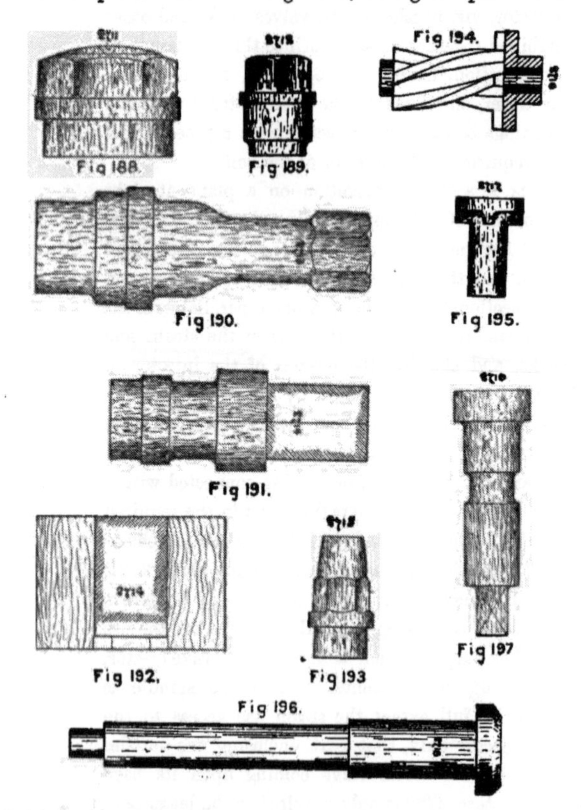
Fig 188. Fig 189. Fig 194. Fig 190. Fig 195. Fig 191. Fig 197 Fig 192. Fig 193 Fig 196.

2723 which closes, when required, communication with the boiler. The back of this valve is also faced, so that when screwed back to its full extent, as it should be when properly at work, it will come against its back seating, and prevent all water and steam passing through the stuffing-box. The casing 2710 receives the steam nozzle or cone 2715, the combining cone 2714, and the delivery cone 2713, all of which are screwed in water-tight joints. The latter, 2713, is pre-

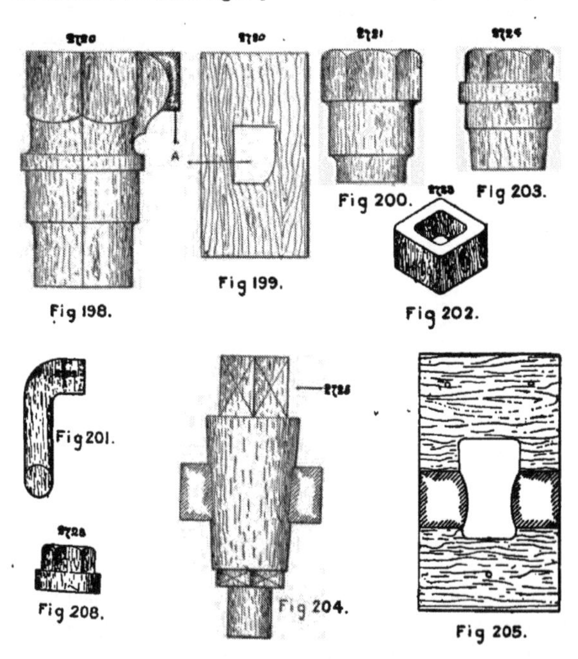

pared to receive the wing cone 2716, or top portion of the combining cone, and in the position shown there is a free outlet for overflow, which passes through the core A, but when the jet is established the overflow ceases. The whole of the cones are inserted by the opening caused by the removal of the injector lid 2711. The steam from the boiler passes through the core C, and the cone 2715, it then meets the water which enters through the core B, at the combining cone 2714, passes through the wing cone 2716, and is delivered by 2713 into the cavity which communicates by the core A with the core D. The back-pressure valve 2717 closes when the jet ceases, and the water is led through the feed portion of the flange into the boiler. Suitable means are devised by which all hot condensed steam from leakage of valves is allowed to drain away as it is formed, thus

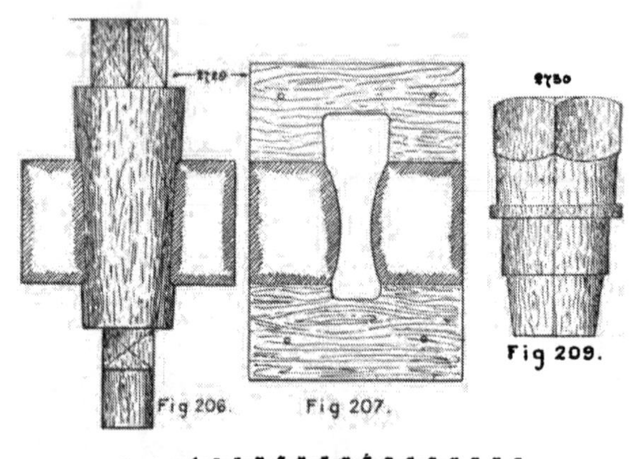

keeping the injector and pipes free from hot water when not at work, so that the injector will start promptly when required. The cap 2712, Fig. 187A, has been removed, and a small push valve substituted, so that in the event of the back pressure valve 2717 sticking, this small push valve can be actuated by the thumb and the valve 2717 released.

In moulding, those portions of the pattern marked B C D are placed upon a turnover board, the flange D being uppermost. The box is then rammed up, a few long nails are placed between C and D to act as gaggers and strengthen the lift, one or two stiff rods being also used right across the

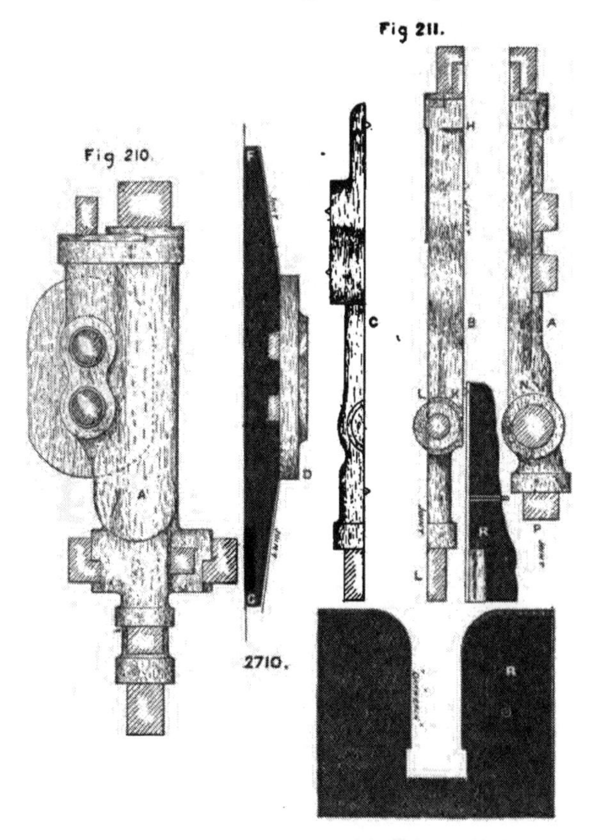

box. The joint F G is then made sloping from the edge of the moulding-box to the edge of the flange, sprinkled with parting sand, rammed up, and a good stiff board placed upon it and turned, the whole mould when finished and ready for casting, resting upon this board. The next joint is made

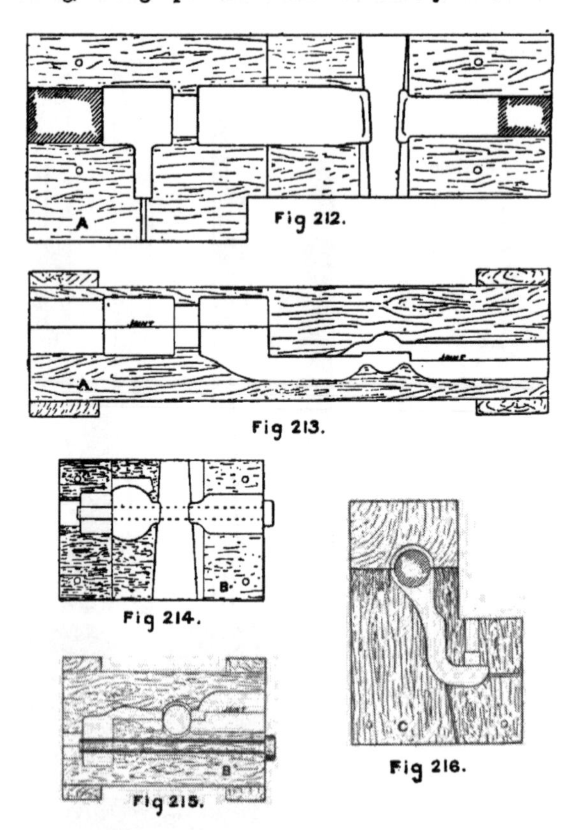

Fig 212.

Fig 213.

Fig 214.

Fig 215.

Fig 216.

along H K L and H N P, the "drawback" R being responsible for that portion of the mould L K N P, the top box being well rammed and vented. The cores, Figs. 212–217, are then placed, every position of which can be well seen and regulated, so that the mould is as perfect as possible, and then closed for casting. If this casting was not moulded as described, it would be necessary to use more drawbacks or print up, the latter being a very objectionable mode of procedure; because then it would be necessary to have a core with its tendency to be set wrong, and also being harder than the mould, it would not give so even a surface all over the casting.

Fig 217.

The number-plates are moulded from a metallic pattern with a system of changeable figures; the cab window-frames, whistle fittings, and the other hundred-and-one items required in a locomotive brass foundry, receive the due amount of attention demanded by their own peculiar construction.

SECTION III.

FORGINGS.

PART I.—FORGE.

ALTHOUGH the manufacture of plates and tires does not come strictly under the head of locomotive building, the forge work would not be complete without some notice, however slight, of such operations. The specification and tests for the plates were given in Section I., the Boiler Shop, and the material stated to be of the best Siemens-Martin mild steel. The method generally adopted in the North of England is to cast ingots from three to five tons in weight, which, after remaining in the moulds from twenty to thirty minutes, are removed into soaking pits, where they radiate their initial heat, or equally distribute it throughout their entire mass for about one and a half hours. They are then cogged down either by hammer or mill, each being equally good, into slabs of suitable size for the mill in which they are to be rolled, or plates they are to form. The cogging mill will either have side rolls of the universal or Belgian type, or an actuating gear, by which the slab is raised or turned about as required, in order to roll it upon the edges. When rolled by the latter arrangement in a vertical position, it is especially beneficial in removing all the scale, it peeling off the flats with the greatest ease. The slabs are then carried forward by means of live rollers to the guillotine, which is actuated either by hydraulic or geared power, having a shearing capacity for slabs up to, say, 3 feet wide by 10 inches thick. The modern arrangements for plate rolling are very complete, the rolls themselves being about 8 feet to 11 feet long by 30 inches and 40 inches in diameter. The labour-saving tools and manipulating gear are so complete in the most modern plate mill in the author's mind, that only four men are required. The reheaters are grouped about the mills in the most convenient manner possible, and every facility is devised for dealing with ingots and slabs, with the most efficient charging and drawing tackle, so that the cost of production is brought down to a minimum.

The section of the engine tire is given in Fig. 218, and the specified deflection for the percussion test is 1½ inch per foot of external diameter, for tires up to and including 3 feet, and 2 inches per foot for all larger sizes. Two per cent. are tested in this manner, selected from bulk, and samples cut from these selections must bear an ultimate tensile strain of not less than 46 tons per square inch, with a minimum elongation of 20 per cent. on 3 inches for Siemens steel. The deflection for Bessemer steel tires must be 2 inches per foot of external diameter, and the ultimate tensile strength not less than 35 tons per square inch, with a minimum elongation of 20 per cent. on 3 inches.

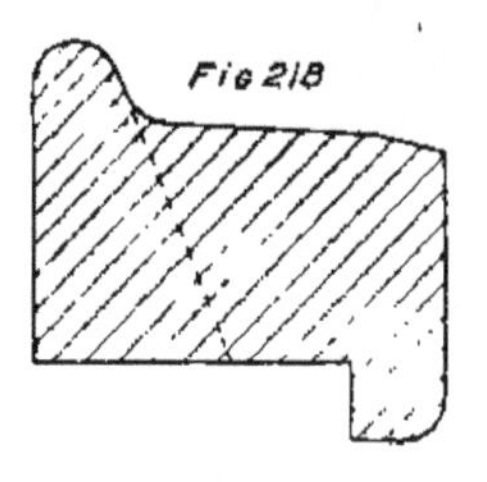

The usual method of manufacture is to cast suitable ingots, hammer them down, punch, beak and roll. The form of ingot is almost of infinite variety, from the ordinary cheese with oval top and bottom and hexagonal or octagonal sides, to that of a Field's night light, each having for its object a good sound blank. The spongy mass and dirt rises to the top, where, after hammering down, it can be punched out. In hammering down, all fins and flashes are cut away, and care is taken that all cracks and traces of ganister are removed. The ingot moulds are kept in good condition, and of course it follows that the greater care there is bestowed on the metal at the melters and during the punching and beaking, the less percentage loss will take place at the mill. After punching, sometimes the blanks are hammered upon the edges or periphery upon a mandril attached to a porter bar. They are then beaked, the larger tires by a beaking recess in the hammer block, the small tires by an ordinary smith's fuller. Some firms do not beak, but have roughing-out mills, and others employ both. Afterwards they are weighed, this being essential, otherwise the percentage of wasters after rolling would be unusually high, through errors in weight, the blanks not rolling to the right section when at the required diameter.

It is very beneficial to have the speed of the mill sufficiently great, to allow a good heat on the tire when finished. Some firms take the precaution to cover their tires immediately with ashes, which gives an annealing effect and certainly preserves them from chills. After rolling, the tires are carefully examined for inside and outside diameter, those too large being rejected, whereas those too small are blocked, and oval tires made round by hydraulic pressure. The two latter operations are considered to be bad practice, and are seldom required by a good roller, but of course when a mill is hard pressed, it is very probable that a few such tires may be rolled, perhaps to the extent of 10 or 12 per cent. This amount is not excessive, as a dozen oval tires can be rectified in about fifteen minutes.

Great care is bestowed upon the selection of the heats for tires of extra quality to satisfy stringent specifications, sample ingots being taken, rolled and annealed until the required tenacity and ductility is obtained, not that mechanical means should supersede chemical in producing the required result.

Forty-five tons tenacity may be considered sufficient for home use, but for India and other hot climates the tenacity may go up another five or six tons per square inch with perfect safety. The material should be as little liable to molecular changes from a repetition of sudden shocks as possible, which points to a tire having considerable ductility, which would fulfil all requirements, except that it would wear out too soon. The rack and pinion theory is well known, in which is advocated the equality of temper in the material both for tires and rails. The author will be content to leave in the hands of the manufacturers the influence which its chemical composition has upon its physical properties.

Iron forgings from puddled bars are generally formed from piles 15 inches wide, 18 inches long, and about 10 to 12 inches high, heated, re-heated and worked until all cinder and slag are removed, and the required section attained. In piling, bars of different dimensions are used for each row, so that all the joints get fairly crossed, and the whole shingled down to either a slab, the average size of which is about 36 by 24 by 3 inches, or into billets. In making a forging of large dimensions a number of slabs are piled together, with a small piece of iron between each, in order to equalise the heat of the whole, so that the centre may be of the same temperature as the outsides. The centre slabs have V pieces cut out in which to insert the staff end, which is driven in with a tup and securely welded. If a very large forging is required, and it is not thought advisable or is inconvenient to make large piles, double faggoting is resorted to. Wrought iron scrap that has been well rattled, carefully piled and well shingled, makes better forgings than puddled bars. Another method is by " laying " or " weighing up " the staff, by which means it is gradually built up at one end to form the forging.

Where there is a great repetition of the same article, stamping by dies or moulds is the most economical and perfect way of forging. For large work the dies are made of mild steel, which gradually acquire a hard skin, but for small forgings they are made of the best steel castings, in such a manner as to release the work in each case with the utmost facility, and generally, forgings are now designed with this point in view. In the case of hydraulic forging, swages or dies may be used of cast iron, which would very soon be broken up by the percussive blows of a hammer. Sometimes two or three dies are required, each bringing the metal nearer the desired shape, the number of dies and the proportion of work they have to perform being a matter of practical judgment. There should be considerable accommodation for the metal to flow equally, the mechanical work performed should be uniform throughout the whole mass, and all fins and surplus metal should be removed before each succeeding die is brought into operation. A knowledge of M. Tresca's investigations on the flow of solids will materially assist in the design and judgment upon stamping dies and forgings, and M. Arbel's performances deserve citing. He forged wrought iron wheels to 8 feet in diameter by a pair of dies, forming the whole wheel complete. It was constructed by building up in the lower die the whole wheel, sections having already been stamped to the required shape. The upper die was fixed in the hammer tup, and after repeated blows, made a well-finished forging. This became a common process in the manufacture of waggon wheels, but has now been superseded by steel castings.

Hydraulic forging gives a maximum strength by forcing the fibre to flow into the required shape, and considerable economy is also entailed by the tool work saved in the machine shop. Soft steel will work more easily than iron, because it is comparatively without grain, and in the case of massive steel forgings the press, with its slow and powerful compression, will supersede the steam hammer.

With the abrupt blows of the latter, a large proportion of the sudden impact is absorbed upon the exterior of the forging, and, comparatively speaking, very little effect is produced upon the interior of the metal, owing to the resistance offered by the *vis inertia* of the mass to the sudden impact of the blow; and should a hammer be used that is really too light for the work it is performing, this surface absorption is so great that the forging or ingot has considerable tendency to pipe, or become hollow and unsound in the centre. In producing large forgings from ingots of mild steel, it is essential that they should receive throughout their whole mass as nearly as possible an equal distribution of the force required to shape them; and with forgings having sharp corners a pressure of 10 tons per square inch is required, and owing to the difficulty in obtaining duplicate work—certainly without repetition it will not pay—and by the rivalry of sound trustworthy castings, considerable impediments are brought to bear against the ardour that might be displayed in developing this class of work.

Crank axles are chiefly manufactured from Siemens-Martin steel, which, after forging and annealing, shall give results upon test bars machined from the blocks removed from between the sweeps, without any further hammering or annealing, of 28 to 32 tons per square inch tenacity, the elongation being not less than 20 per cent. on 3 inches, Whitworth or Woolwich bars; and a 1¼ inch bending test, placed upon 6 inch supports, to obtain a right angle without any signs of failure. The first results upon opening the test records of cranks made to the above specification for the Lancashire and Yorkshire Railway are 29 to 31 tons per square inch tenacity, 27 to 34 per cent. elongation, 39 to 50 per cent. contraction of area, and the bends doubling flat upon themselves.

An ingot should be cast considerably heavier than the

finished forging, in order to get sufficient material to form a sound crank, the bottom portion of the ingot of course being used, and generally, when roughed down under a vertical hammer, this extra weight forms the bottle-neck in that class of ingot, and this in its turn forms an attachment for clamping the heavy porter bar.

It is contended by most crank makers that the ingot should be about 24 inches square, and weigh not less than 70 cwt.; at least, it should be heavy enough to allow one-third of the upper portion to be removed, owing to the tendency of very mild steel to pipe. After casting the ingot, this pipe should be allowed sufficient time to definitely form itself, otherwise if the ingot is turned over or upon its side,

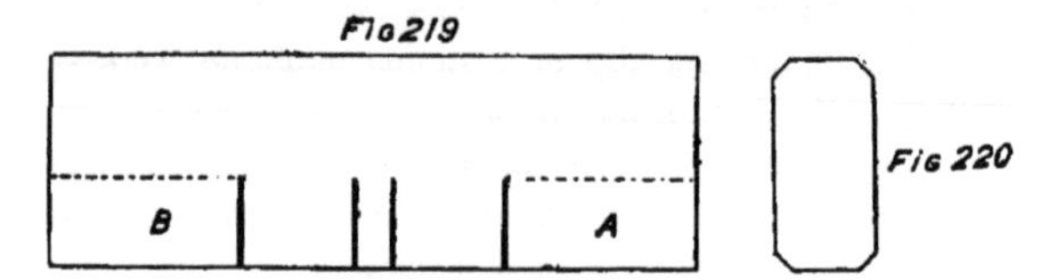

part of the pipe will be parallel to the axis and part at right angles, or in some other objectionable place, and consequently it will be found later on somewhere in the forging. It is advisable to charge them into the re-heaters, or better, into the soaking pits before cooling; because in cooling sometimes a clink occurs, which is not always external, and consequently invisible even after machining; that is, an internal rupture takes place which will eventually become the starting place of a growing flaw. It has been found that steels high in manganese and silicon are especially prone to this evil. The hammer used for cogging the ingot down into a slab, Figs. 219 and 220, should not be less than 15 tons, but a smaller, about eight tons, may be used for finishing. Care should be exercised in heating, as a repetition without mechanical work will change the nature of the fracture of any forging.

In the present case the ingot is drawn down by a 35-ton duplex hammer to a slab 24½ inches by 12 inches, Figs. 219 and 220, and immediately brought to the hot saw, and sawn as indicated by Fig. 219, for the sweeps. It is finished at an 8-ton vertical hammer, the first operation being to cut out the gusset A, Fig. 219, with a large cutter, and then roughly round up this end to 9½ inches diameter. The portion between the sweeps is then removed by cutters, the staff changed, re-heated, and the gusset B, Fig. 219, cut out, and this end roughly rounded to 9½ inches diameter. The hammer blocks are then changed for a narrower set, so that the middle can be rounded roughly, and then re-heated for twisting, right-hand sweep to lead. The sweep that follows is then placed between the anvil and tup, and the leading sweep has a large spanner attached. It is then drawn up at right angles by suitable means, utilising hydraulic or any other available power.

For obvious reasons, the leading sweep is the one always twisted, as it will then always precede, whether in forward or backward gear. The middle is then hammered to "peg," the finished forge size being 8 inches diameter, and the blocks changed for the ordinary set. The ends are then finished exactly to 9½ inches diameter, forge size, utilising a peg, and cut to length, each end resting in turn upon a swage to preserve its symmetry, and ordinary cutting tools are used. The sweeps are then rounded under the hammer, top and bottom, and if found to be of slightly insufficient width, a fuller, consisting of a round bar, is stamped down the centre, which widens them out, without affecting the thickness where it is required. The finished forging is shown in Figs. 221 and 222, but minus this groove.

It may be here observed that all the following forgings

are manufactured from a mild steel of similar quality to that of crank axles, the exceptions being noted where necessary. In every case their utility and durability depend to a greater extent upon their ductility, than a high tenacity. They are subjected to such stress and strain in everyday working, that a high tenacity would be absolutely detrimental to an

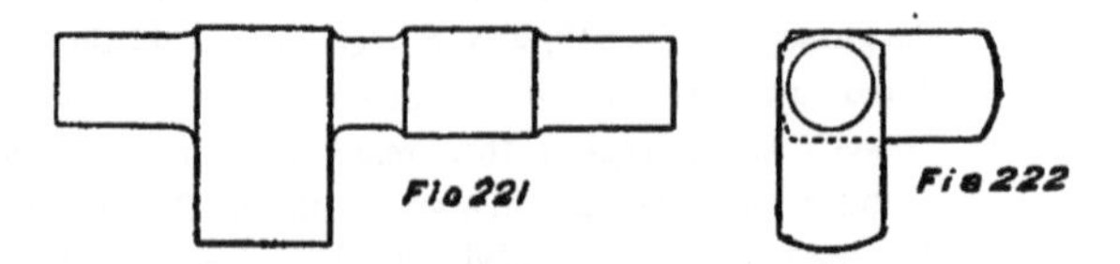

economical life, and to the safety of the travelling public. When necessary they are lined with case-hardened bushes or faces, that is, where great wear and tear take place locally.

The crank pins for the outside rods are forged as indicated by Figs. 223 and 224; three or four continuously, that is, in one length of bar, as shown, and afterwards sawn off. Blooms

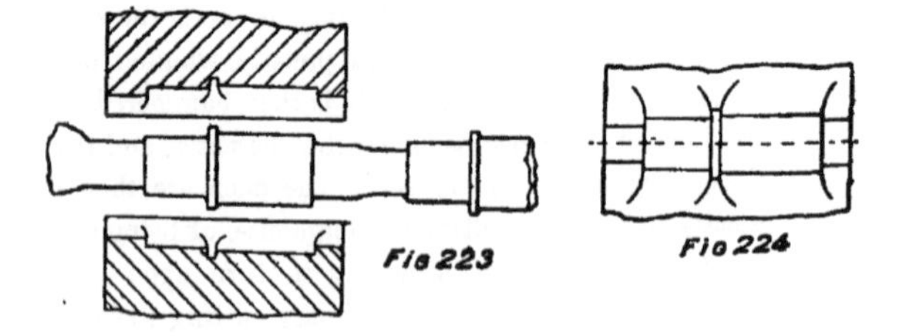

are roughed out to 5 inches square, and about 3 feet 6 inches long, and the pins are swaged out to the required size by a pair of fast hammer blocks, no other tools being required. The washers for these pins are stamped in a loose die, from round bars 3 inches diameter cut to a length of 6 inches, the hole being punched through a guide plate, Fig. 225.

The specified tensile tests for engine straight axles are the same as for crank axles, besides which they must withstand sixteen blows from a 1-ton tup falling 25 feet, resting upon centres 3 feet 9 inches apart, after each blow the axle being turned and the test continued until fracture, generally requiring to be nicked. The deflection will of course fluctuate according to the temper of the material, but as the percentage of carbon will only vary a few points—hundredths per cent.—it follows that the deflection is generally found to be fairly constant, that for engine axles being about 1½ to 1¾ inch, and tender axles from 3 to 5 inches for each blow, and becoming straight, or nearly so, upon reversal. The

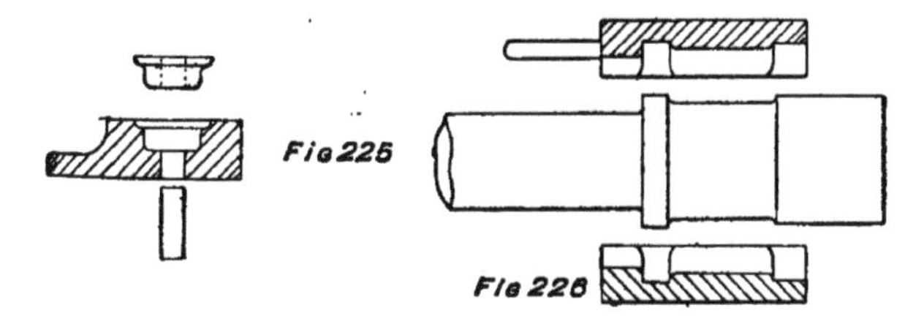

heat developed in these tests is very considerable, but the author has known an engine axle with a deflection of nearly 5 inches to be straightened by blows from the tup falling 25 feet, after remaining in its deformed condition for four days, without being annealed. Within the specified tenacity, 26 and 28 per cent. elongation is obtained both on the length and crossway samples, and from 40 to 60 per cent. contraction of area. They are made from blooms 4 feet 9 inches long, 9½ inches square, with round corners, at a 5-ton hammer. One end is rounded to a peg 9 inches in diameter, and cut off in swages, then the neck and collar are stamped in a pair of swages as indicated by Fig. 226, the middle being hammered to peg, leaving about ¼ inch all over for turning.

In some works the blooms are rolled to the rough section here shown in a cogging mill, and in all cases the axles are

swaged at two heats, one half of each axle being finished before the other half is commenced, either by the swages as shown, or fast hammer blocks, this being the most rapid and economical way of forging them. The coupling rods are forged from a slab 12 by 4½ inches by 3 feet 1 inch, the middle being first drawn down to pegs 4¾ by 2 inches, then the bosses for the pins are stamped in a pair of swages, Fig. 227, the whole job being

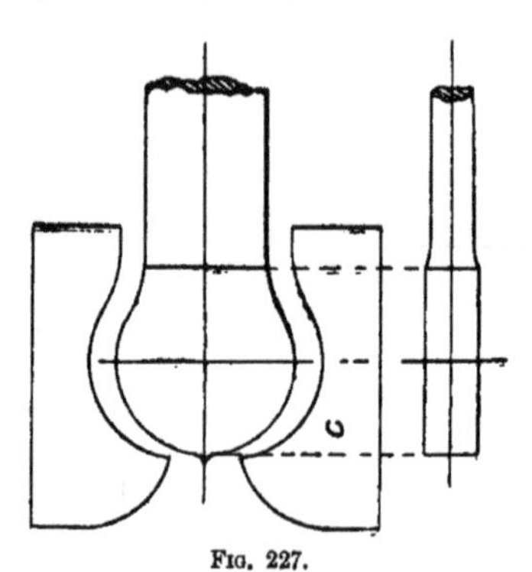

FIG. 227.

finished and set for the machine shop without any further trimming up in the smithy. The following sketches, Fig. 227A, clearly show the operation of punching these rods: F is a plan of the block D, which is a steel casting, and G is a sectional elevation upon A B, also giving another view of the rod, shown in Fig. 227. The rod is forged slightly smaller than the block and a little thicker than required, the latter being of exact forge size, so that during punching the forging swells, although slightly drawn in at the top, and fits the block exactly; and this necessitates the loose block E to facilitate the removal of the rod, which operation is obvious. The connecting-rods are forged in pairs; that is, two big

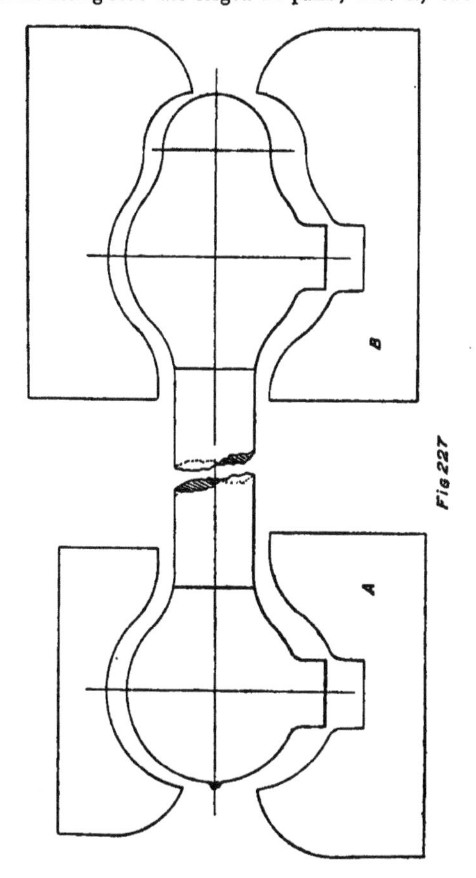

Fig 227

ends are forged in one piece, and after the rods are finished they are divided at the big ends, the tools used being simply the peg, plate, and the swages for the enlargement for the Joy motion pin. After hammering down the big ends to peg, the radius A, Figs. 228 and 229, is formed by the hammer blocks.

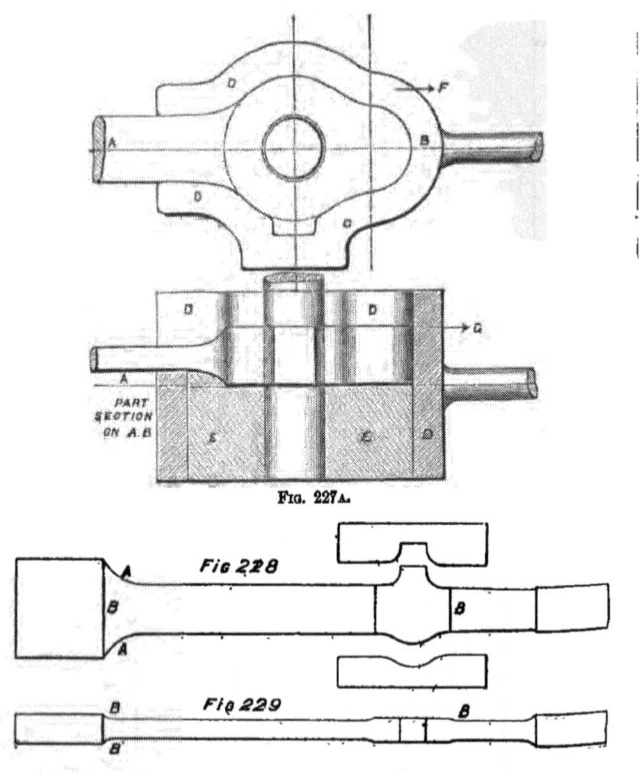

FIG. 227A.

All the hammer blocks have a radius of about two inches at one edge, this being used for all forgings requiring a similar radius or larger; but if smaller, it is set in with a bar. The middles are drawn down to a peg, and the radius B, Fig. 229, is formed by plates. Fig. 228 shows the swages for the enlargement in the centre of the rod for the oil cup, and after the small ends have been drawn down to a peg, the rods are cut to length and sent from the forge ready for the milling machine. Fig. 229A shows the forging for the big end strap, and Fig. 230 in its finished state. A bloom is roughed out to 7 by 4 inches by 3 feet 7 inches, making two straps, one end of which is drawn down to $5\frac{1}{8}$ by $3\frac{1}{8}$ inches for about 20 inches, and the rest to $3\frac{1}{8}$ inches square. The end $5\frac{1}{8}$ inches deep is then set down by a tool, as shown in Fig. 229B, to $3\frac{1}{8}$ inch square, leaving the oil cup as required, and afterwards the knees B, Fig. 229A, are set in

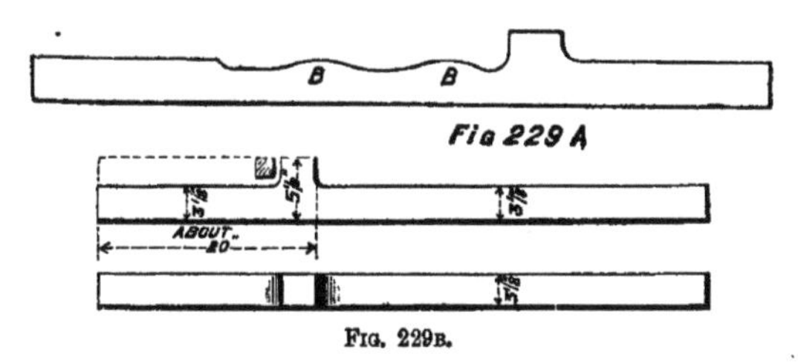

FIG. 229B.

with a fuller for the corners in bending, which operation is performed in the smithy, making use of the block, Fig. 229C. A is a cross-section, and F a plan of this, and it is practically a substitute for the hammer block. The small curved plate B is placed at the bottom of the strap recess at D resting upon C, the function of which is to give the strap good round corners and leave plenty of metal for the smith to work up afterwards, whereas C is simply a convenience for raising or starting the strap upwards by means of a couple of wedges, if it has been wedged by the bending process. The actual bending is accomplished by means of a few blows from the hammer, transmitted through the rectangular block E, and one or two making-up pieces.

The little end straps, Figs. 231 and 232, are made in a similar manner from a solid bloom about 6 inches square, drawn down to about 4 inches square, after which the oil cup

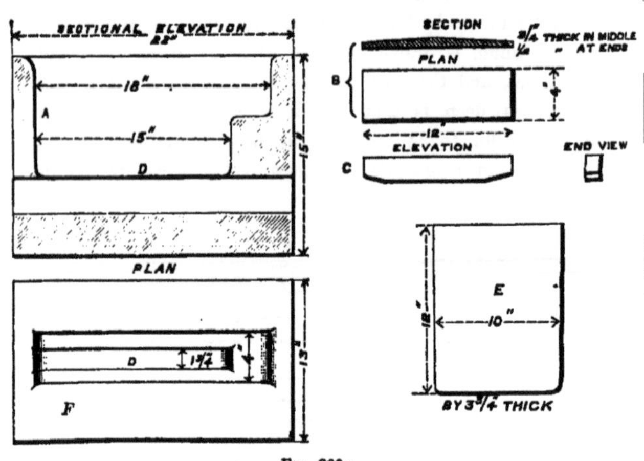

FIG. 229c.

is struck out by a tool having the following section, and the rest of the strap is drawn down upon each side of the cup to a peg $3\frac{1}{4}$ by $2\frac{1}{2}$ and 2 inches. They are

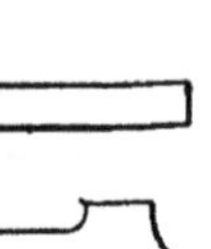

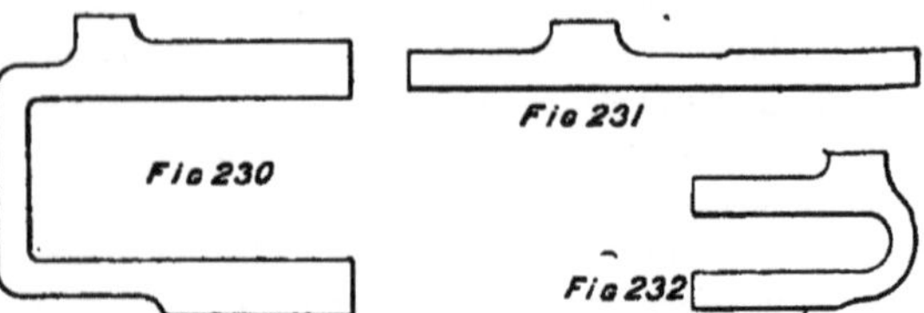

then sent to the smithy in the form of Fig. 231 and afterwards bent. Failing more appropriate tools, or lack of quantity, the sketches Fig. 232A, represent the progress of this operation. The U-piece of iron A is about 2½ or 3 inches in diameter, and simply bent to a semicircle, and the forging is placed successively as represented, receiving the blows from the hammer through a fuller as directed by the arrows. After the third representation the forging is ended and receives the blows direct, until it is bent over to nearly the requisite width, when it is finished exactly by placing a block inside

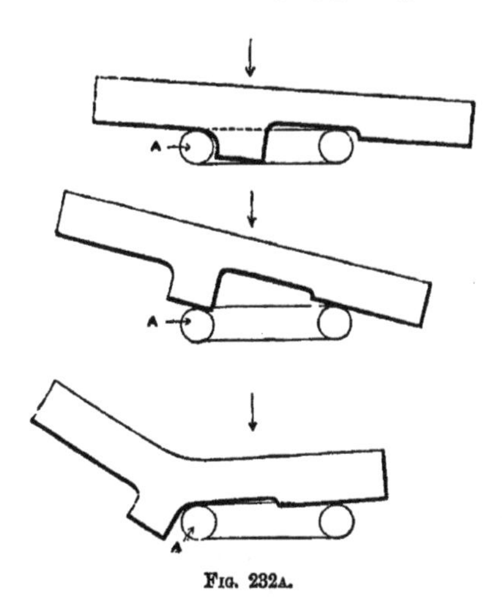

FIG. 232A.

the bend, and then closing by a few more blows, after which it is hammered to thickness upon its edges and pared up.

The valve rods and anchor links for Joy's valve motion are made in a similar manner to each other, and are shown in Figs. 233 and 234, and consequently one description of the manufacture will suffice for both, the example given being the valve rod, the only difference being in the length A. A bloom is roughed out to 7 by 4 inches, and then drawn down to 5¼ by 3¾ inches, the largest dimensions of the ends, after which the middle is drawn down to peg 2¼ by 1¼ inches. Swages, Fig. 233, are used to form the shape of the ends, after

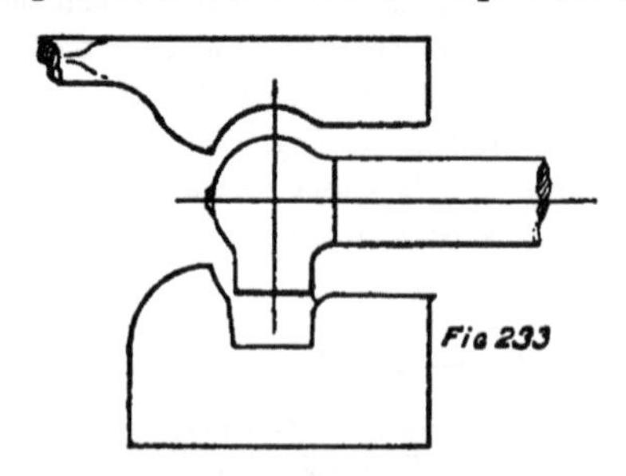
Fig 233

which they are worked in the die, Fig. 235, in section, a plan of outline being given in dotted lines in Fig. 234, which leaves

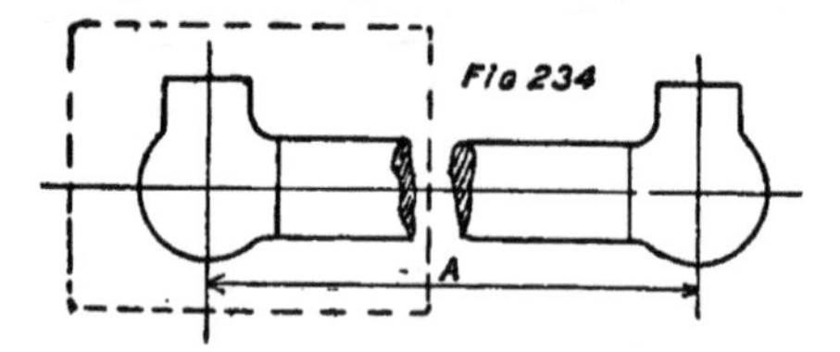

Fig 234

them a well-finished job, ready for the machine shop. The radius links for the slipper blocks for Joy's motion, which can be clearly surmised from the blocks used, are made from solid best iron blooms about 7 by 5½ inches by 1 foot 7 inches, requiring the group of tools shown in Fig. 236, when a restricted quantity of links suffice the necessities of the shop, and when it would not pay to roll them in a merchant mill and set them to the required radius afterwards. After re-heating,

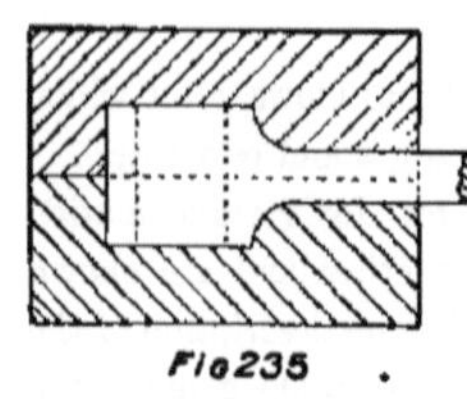
Fig 235

the first operation is to stamp a straight rectangular bar into the bloom, making it into a channel, and then follow on with

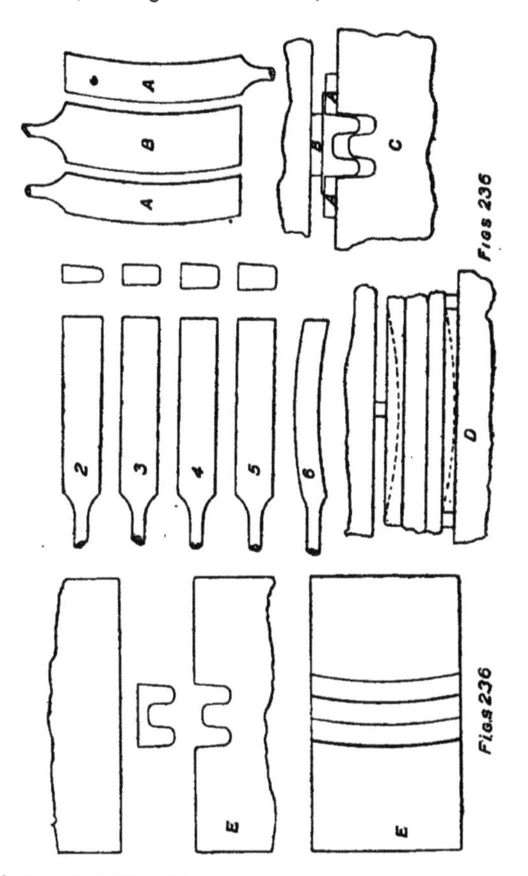

Figs 236

Figs 236

the bars 2–5, Fig. 236, which gradually open out the groove. Then this rough forging is bent to radius as indicated by D,

Fig. 236, and afterwards the bar 6 is inserted, having the exact radius. The forging is then re-heated and stamped into the blocks E, and afterwards the fins and flashes removed

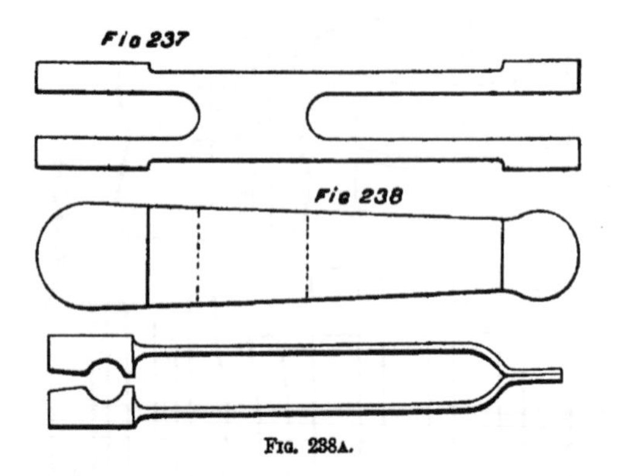

Fig. 238A.

as indicated by C, Fig. 236, then sawn to $16\frac{3}{8}$ inches long ready for the machine shop.

The stirrup link, Figs. 237 and 238, is at present chiefly smithy work, and is produced from a forging 4 by 3 by

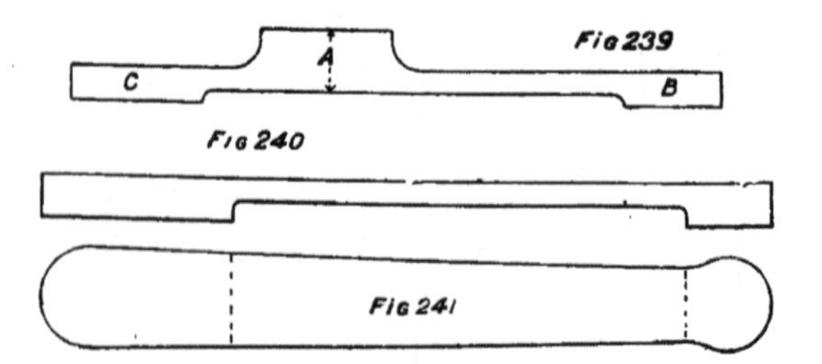

14 inches, being first of all smithed out to Fig. 239; A being about $2\frac{1}{2}$ inches high, B to finish forge dimensions, utilising the clapper swages, Fig. 238A. Two similar forgings

are prepared, then firmly welded together, and afterwards dressed up. The swing links, Figs. 240 and 241, are worked from forgings $4\frac{1}{4}$ by $10\frac{1}{2}$ by 20 inches, being first set down in the middle and pared off or finished in clapper swages. The valve buckles, Fig. 242, are forged from wrought-iron piles of about 2 cwt. in weight, which are shingled at a welding heat into blooms about 7 by 3 inches. These blooms are re-heated and roughed out under the hammer, being

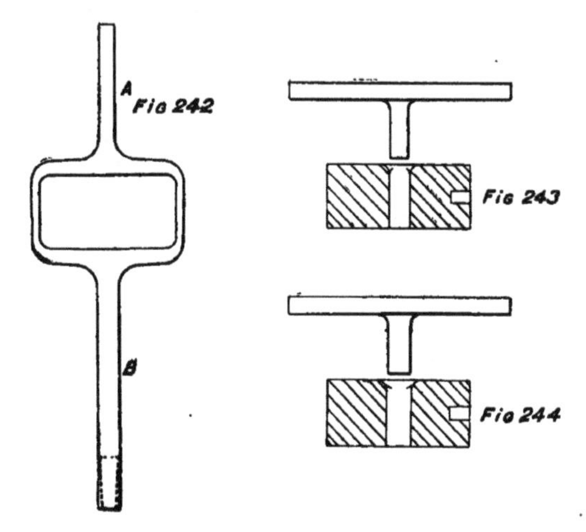

drawn down by the aid of plating tools, and the shanks are roughly rounded to peg. Afterwards the ends A and B, Fig. 242, which is the finished forging, are stamped in bored blocks $1\frac{3}{4}$ inch for the dummy gland end, Fig. 243, and $2\frac{7}{16}$ by 6 inches long for the welding or spindle end, Fig. 244, which jumps them to fill the hole, leaving $\frac{1}{4}$ inch on for turning at the short end, and ample metal for scarfing and welding at the opposite end. They are made in pairs in the forge, and finished in the smithy. They are received by the smith generally as in Fig. 244A, but occasionally, when the forge is hard pressed, one or two may be sent as indicated by the dotted lines. The first operation then, is to draw out the ends under the hammer, and at the same heat stamp them again in the blocks which set the taper. Afterwards the ends are again thinned by a plating tool. Each portion is bent by the aid of the tool A, Fig. 244B, the shank B fitting into the square hole at the flat end of the anvil. After each portion has been bent it is scarfed for

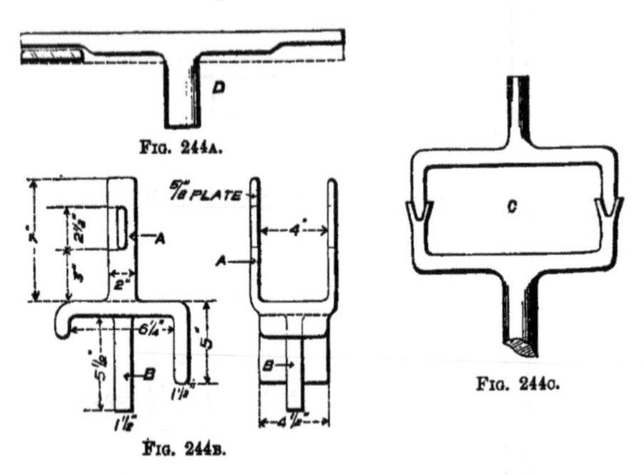

FIG. 244A.

FIG. 244B.

FIG. 244C.

welding, Fig. 244C. The spindle is then welded on and the whole set to template, ready for the machine shop.

When a solid cross-head is not used, but a steel casting, as shown in Figs. 89 and 90, p. 89, or when a shank is not cast to the cross-head and drawn down, Fig. 245 shows the hammer blocks used in forging an ordinary piston-rod, which shows clearly the mode of manufacture. Fig. 246 gives two views of the spring buckles, the first forging and its block being shown in Fig. 247. They are worked out of a piece of best iron 8 inches square, then set down upon each side of the boss for the pin until prepared for the block, in which they are then stamped. Afterwards the flats are drawn out, bent and welded up. The only forge work that now remains is the slide bars, which are roughed out 4⅛ inches square and 2 feet long, then re-heated and rolled in a 14-inch merchant

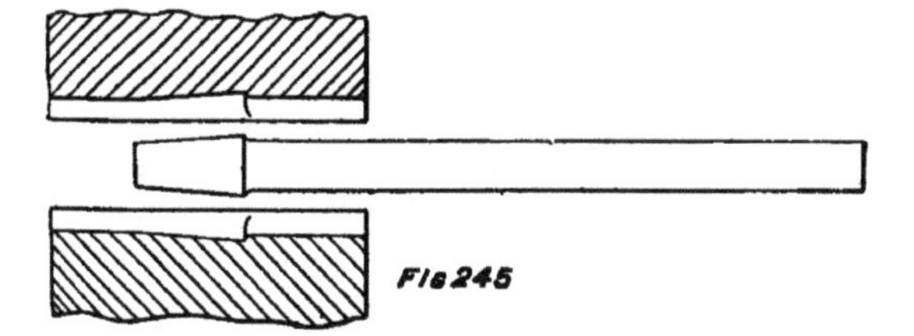
Fig 245

mill to 3¼ by 2¼ inches and cut to 4 feet 3 inches long. They are then set under a hammer after being straightened whilst hot on the floor plates, ready for the milling machine.

The whole of the forgings described, are made and finished in such a manner that there shall be a minimum amount of finishing tool work required, consequently entailing a

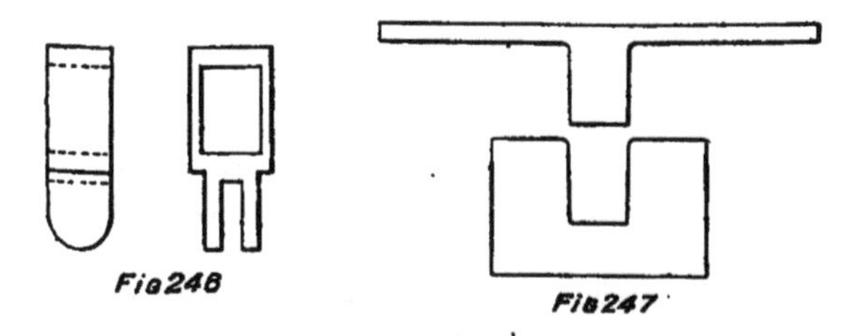
Fig 246 Fig 247

maximum despatch in the machine shop and reducing the cost. They are also stamped with distinctive marks, which are carried forward in the machine and other shops. This system is almost universal, so that in the event of failure of a tire, crank, rod, or any other piece of mechanism its history can

can be clearly traced. Moreover, the collection of templates and gauges is constantly kept up to standard, and all forgings worked to these templates. Mandrils and finished forgings are kept purposely for striking swages from, so that indifferent work shall not be possible through lack of convenience in replacing damaged blocks. One mandril may be made up, representing two or more forms of straight axles thus, Fig. 247A, showing a three-standard mandril, each section being turned up, whilst another may have the crank pin ends of the coupling rods. After the swage has been struck, a hole is generally punched at the end or side, and a bar from ½ to 1¼ inch diameter inserted, which is gripped by

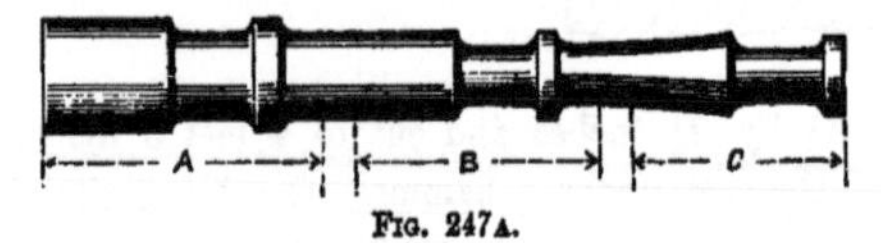

FIG. 247A.

the cooling and contracting swage, or pieces to form handles are sometimes dabbed on, and afterwards, if it is necessary, the impression formed by striking the swage is finished by a fitter chipping it up, or otherwise perfecting. If the swage is to be of the clapper type the bows are simply welded on; and finally, fixed blocks are more conducive to good and rapid working than loose ones. An example of this class may be taken in the coupling rods, Fig. 227, one end of the block having that portion marked A, whilst the opposite end swages the centre joint of the rod B, with a good true surface in between for drawing down the rod, having the required radius at the edges. The work is also carefully portioned out, each hammer having essentially nothing beyond its power to deal with, for reasons already put forth earlier in this section, and also retaining its set of men, shingler, furnacemen and helpers, for division of labour is to a great extent answerable for the successful working of a forge.

Part II.—Smithy, including Springs.

Undoubtedly the smith's art is of great antiquity, and it is sometimes a question why the early artificers in metals should have blended copper and tin, during the bronze period, before creating what the author might term a welding period. Doubtless the requirement of a high temperature, created by a blast, was the impediment, for it is related that one of the earliest productions of this very necessary accessory to the smithy was the selection of a valley, with known prevailing winds, then boring two or more cone-shaped tunnels through a hill, with the bell mouth to the wind, and tapering them to meet in one narrow neck. The ancient smith also appears to have been a factor of great power among the armies of his day, for upon him the conquering made special raids, in order that the vanquished should be without the means of reproducing those wonderful blades, which cut that which offered no resistance, and consequently remain for a longer period under submission. The author is fully aware that this is probably of more interest from an archæological point of view than mechanical, and he offers the most beautiful and wonderful work of the early English smith, such as the fabrication of most costly armour and the production of ecclesiastical ornamental work, as his apology for introducing such remarks to his readers.

It is somewhat difficult to draw a definite line of demarcation between the forge and smithy work, consequently the section devoted to the forge should be read conjointly with the present one, because in many cases the former has to prepare rough forgings for the latter, and active management is always working in the direction of saving as much of this class of work as possible, consequently the larger and heavier portions of smithy work are constantly being devised—by the aid of more complete tools—to be finished in the forge for the machine shop; therefore for this reason and the use of steel castings, Section II., Part II., page 72, the smith's work is being periodically curtailed.

Wrought iron is the material upon which the smith is generally engaged, but mild steel, both acid and basic, is now being extensively used, and, after a little experience, it is a fact that a smith will prefer the latter, especially if the former has been produced from inferior scrap, or from scrap that has been worked a number of times, or is red-short. The relative qualities of the various brands of wrought iron may be pretty accurately gauged by a comparison of fractures, provided they have all been broken under similar conditions, a ready means being to nick and bend across the anvil. It is obvious that for comparison the conditions must be similar, because the fracture produced by a tensile strain will be vastly different from that produced by bending. The texture of good wrought iron is fine and close, and of a silver grey colour. It is well exemplified by nicking a 1¼-inch bar about ⅛-inch deep, and then bending the sample down flat upon itself, when the fibre will be found to be quite distinct and long, whereas inferior iron may be coarse and granular, with short fibre of a dark colour, indicating cold-shortness, and sometimes largely crystalline. It is certain that the element of time will affect the fracture, and to bring out the fibre thoroughly, the bending operation should be performed slowly and gradually.

The principal operation with which the smith is concerned is welding, which gives to iron its great intrinsic value, and is generally considered to be an adhesion under pressure of two pieces, which have been heated up to the plastic limit. Practically the two are made into one, but the great difficulty in removing dirt, scoria, oxide or other foreign matter, renders an actual and perfect contact almost impossible;

and however good a weld may be, and whatever attention may have been bestowed upon its production, a line of demarcation may generally be detected. A fine metallic surface is of course the first requisite, followed immediately, and of equal importance, by a consideration of the impurities incorporated in the iron. These impurities will affect the weld by their own weldability, and their tendency to crystallisation, also if prone to oxidation to a greater degree than the iron in which they are, they must naturally be taken as detrimental. Their weldability may be dispensed with, by stating that there are only two impurities, viz. nickel and cobalt, which will weld, consequently all others must be taken as hurtful. The property of retaining an amorphous or plastic condition up to a high temperature, say approaching its melting point, is favourable to good welding; therefore, anything that tends to lower this condition, such as carbon, silicon and phosphorus, causes the resulting weld to be of an inferior quality, generally known as being cold-short. These impurities are not actually detrimental to the act of welding, but lower the welding temperature, especially in the case of phosphorus. Sulphur behaves differently, it aids the separation out of the more fusible compounds; consequently, iron containing it crumbles when struck. Manganese in high percentages causes the material to be very red-short, and the greatest care is required both in the gradual heating and ultimate manipulation; nothing but the lowest temperatures will permit of the working of a material high in manganese in any degree. Therefore, to produce a sound weld, there must be the fullest contact between the unoxidised metallic surfaces of the metal; and the more homogeneous and pure the material is, the more perfect will be the weld. It is difficult without hammering to bring these surfaces into sufficiently near contact, and also to remove the slag which has been developed to prevent the oxidation of the scarf whilst heating. It is more difficult to weld mild steel or ingot iron, because it requires a quick fire and smart handling, success depending to a greater extent upon the mechanical treatment than chemical composition. It will not weld nearly so well as iron containing a greater percentage of impurities, because in the latter they partake more the nature of an admixture than a chemical combination; and it is found from considerable experience that the welding heat for mild steel is at the point of transition from a bright red to a white heat, which appears to be the temperature, when, as it were, the cohesive state is changed for that of the plastic.

The method of scarfing is an important factor in producing a sound weld, for upon it depends the ease with which the slag is forced out when contact is made. In welding up, say, 2-inch bars, the most frequent method is for the striker and helper, with backing hammers, to set back the bar thus, and afterwards extend it by fullering. As long

as the smith does not proceed any further this will produce a good weld, but frequently he gives the bar a quarter turn and a few blows, the result being to create a hollow, consequently trapping the slag, when the two scarfs are brought together. The best method of scarfing is to use the flatter after the fuller, then bend slightly over the beak of the anvil, which produces a scarf thus, and when two similar scarfs are brought together,

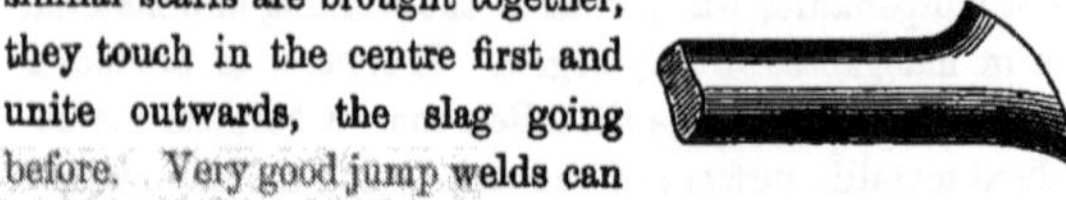

they touch in the centre first and unite outwards, the slag going before. Very good jump welds can be produced, but this method in an ordinary smith's fire is not to be encouraged, because they always break at the joint, showing a crystalline fracture with short elongation, whereas with round scarfs, they break as frequently outside the weld with fibrous fracture, which is also the case when broken at the weld. In good welds it is quite within reason to expect under a tensile strain that the bar will not break at the weld, because the extra mechanical work put into that portion raises its ultimate strength, consequently after fracture the contraction at the weld is less than along the rest of the bar, but taking the whole shop through, it is not safe to calculate the strength of a weld beyond 70 per cent. of the solid.

It is not proposed to deal with the many mixtures used as fluxes, especially in the welding of mild steel, but it is only fair to state that there are certain compositions on the market which are valuable accessories in welding high carbon and tool steels. They cannot exactly, however, be wholly passed over, because certain materials, such as sand, are commonly used, and are useful both for iron and steel. The metal is inevitably oxidised during heating. Wrought iron and soft ingot iron or mild steel may be safely heated to the temperature at which the oxide is fluid, then a flux is not necessary, but might facilitate welding. Again, as the percentage content of carbon is increased, the welding point is correspondingly decreased, therefore a flux must be used to slag out the oxide at this lower temperature. From this the reason is readily surmised for the numerous fluxes or welding compositions upon the market, which are, generally speaking, alkaline salts. It must be remembered that the flux does not benefit the metal, and the chief requisites are low melting points and cheapness; therefore, as ordinary sand and borax admirably fulfil these requirements, they are not likely to be readily supplanted. They form a slag which simply surrounds the parts to be welded, and protects them from oxidation and the influences of the impurities in the coke, especially sulphur. Impure iron will form a flux out of its own impurities, and when the scarfs are brought together and hammered or otherwise compressed, the slag is squeezed out.

Although the chief requirements for welding mild steel have already been stated, sometimes calcined borax, mixed with salammoniac, is used, but there is no apparent necessity for this, although, certainly it makes a more fusible flux.

Welding by electricity having been made the subject of considerable commercial enterprise, it will not be entirely without interest in a locomotive smithy, and a short notice of two systems—the low potential and the arc—may be of service for special purposes rather than general use. In both systems there is economy of time and labour, a weld being performed with the utmost rapidity at a very low figure, for as many as 700 welds, ranging from $\frac{3}{4}$-inch bars to 2-inch shafting, can be made per week by a 40 A type machine of The Electric Welding Company, Limited. There is also saving in material, for there is no blistering, scale or burning, and if two pieces, each 1 foot long, are placed in the machine, when finished the product will be one bar exactly 2 feet long. The heat can be regulated perfectly, being always visible, and there is not any necessity to allow 1 inch or $1\frac{1}{2}$ inches for scarfing. The absence of scale and dirt in the arc system or improved Benardo's, as carried out by Messrs. Lloyd and Lloyd, Birmingham, is probably due to the fact that the vapour of iron is produced to such an extent, that the atmosphere cannot get sufficiently near to oxidise the metal. Both systems have great advantages in welding uneven sections, and also prevent the introduction of dirt, whether solid or gaseous, and should two bars be brought together, the ends to be welded being covered with oxide, it does not appear to be the least detrimental, it escaping as a drop or two of slag falling from the surfaces when under pressure. The principle of the first process is the conversion of a current of very high potential, by means of transformers, into low potential. Machines are designed and supplied to suit the nature of the work to be done, whether welding or brazing, and to utilise whatever existing power may be at hand, and each machine is capable of welding various sizes within reasonable limits. The current of low potential is caused to pass through the abutting ends of the pieces of metal to be welded, rounds, squares, angles, tees, zeds, &c., generating heat at the junction, which is the place of greatest resistance. This machine has also been found very useful in welding up short lengths of turning, planing and other tools—Mushet—or otherwise, and thereby increasing their life. Union is effected by following up the softening metal by mechanical means at a welding temperature. The weld in this case is as efficient in the centre as the exterior, because the heat generated by the current commences at the centre and extends outwards; but should there be a point of first contact elsewhere, of course the heat is first generated there. Generally, a burr, upsetting or enlargment of the section is produced, which is either removed by a light hammer or by specially designed machinery. Mechanical work is of course beneficial to the weld, consequently an anvil is mostly found in line with the clamps of the machine, with a part section of the article being welded, let into its face to facilitate hammering up. Good illustrations of this work are the welding on of new ends to old iron and steel boiler tubes, in which case they are hammered on a mandril; and also the upsetting of tubes or bars at any position of their length.

In the arc system an ordinary lighting low tension continuous dynamo is used, in conjuction with a system of accumulators, which come into action only when the actual welding is taking place. The reason for the latter is to get a high efficiency from the plant, the accumulators acting the same part as an accumulator in relation to the pumps in a system of hydraulic pressure. One terminal is connected by means of a flexible cable to a carbon, held in an insulated holder by the workman. The other terminal is connected to the table on which the work lies, or to the work itself, and the arc is sprung between these poles by the workman touching the work with the carbon, and then raising it up. The arc should be as long as possible for good regular work, the maximum being for welding purposes about 6 inches in length, and having a sectional area of about 2 square inches. Iron and steel are made the positive poles, and as this is the pole that volatilises, the vapour of the metal will generally assist in maintaining the arc, but for other metals the poles are sometimes reversed. To avoid any great concentration of heat on a confined area, the carbon is caused to vibrate at a high rate of speed by mechanical means, actuated by a small electric motor, as is also a small hammer, used for securing welds and removing burrs. The mechanical motion of the carbon is produced by fixing it eccentrically to a spindle, which revolves at about 300 to 400 revolutions per minute, the circle produced being about 2½ inches diameter, and this spindle has also a longitudinal movement of about 4 inches at the rate of thirty strokes per minute. The degree of heat is regulated to suit the nature of the work, and ordinary welds are generally produced by burning on small pieces of iron and steel, as the case may be, weighing only a few ounces. Owing to the intense heat and light of this system, the eyes and face of the workman must be protected, and it has this disadvantage, it lacks a ready means of control over the temperature, which is extremely localised, much more so than in an ordinary smith's fire; consequently the internal strains set up must be objectionable, and the high temperature developed causes a greater amount of local crystallisation. This system is less desirable when the sectional area is small, but where there is greater cubical content, such as internal flues of boilers and pipes, it is more convenient than the low potential.

Angle smithy.—The hydraulic flanging press has considerably curtailed the work of the angle smith, but nevertheless there is much interesting work being done, as the author will endeavour to show. The anvil block in general use for angle work is shown in Fig. 248, which will answer for most purposes, but where there is a repetition of an acute or obtuse angle, a block is made having the required angle, and also there are blocks with segments of circles upon each edge, with various radii for angle-ring making. The general process may be illustrated simply by describing the work and manner of making a right angle. The angle iron is marked to the required length in each direction, A B, then upon each side of the vertical an angle of 45° is marked, that is, nicked cold and afterwards heated and removed by a hot set. The edges are then scarfed, and the angle bent round. Sometimes the corner is hammered square, and at others a "glut," that is, a small length of bar iron, is welded on. The angle to which it is first bent is an important consideration and perhaps, to be on the safe side, it should be rather more acute than is required at the finish, but experienced journeymen appear to work correctly without this preliminary operation. However, if it is rather more acute, and the scarf rather thicker than the rest of the section, the angle can be easily obtained by hammering the joint a little more, and then if too thick, metal can always be removed by the hot set when it would be impossible to add

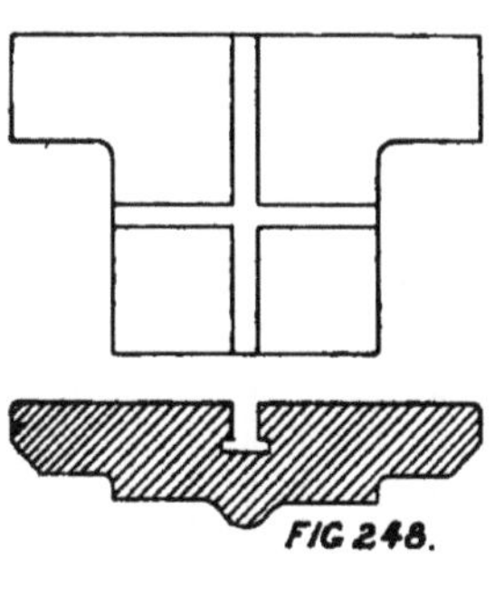

FIG. 248.

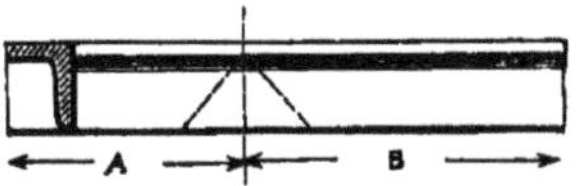

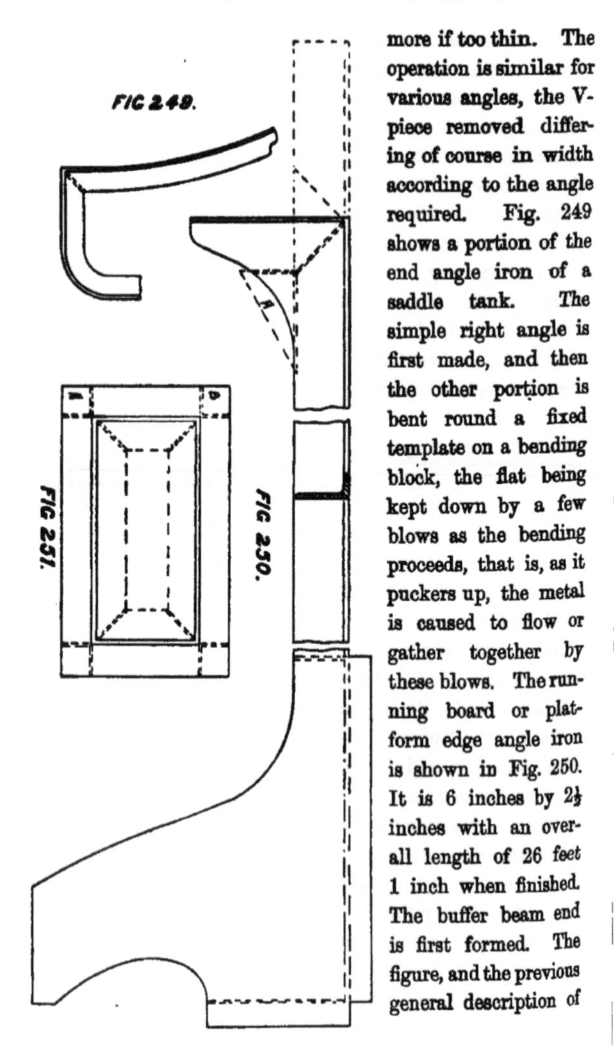

more if too thin. The operation is similar for various angles, the V-piece removed differing of course in width according to the angle required. Fig. 249 shows a portion of the end angle iron of a saddle tank. The simple right angle is first made, and then the other portion is bent round a fixed template on a bending block, the flat being kept down by a few blows as the bending proceeds, that is, as it puckers up, the metal is caused to flow or gather together by these blows. The running board or platform edge angle iron is shown in Fig. 250. It is 6 inches by 2½ inches with an overall length of 26 feet 1 inch when finished. The buffer beam end is first formed. The figure, and the previous general description of making a right angle, clearly indicate the process, and is followed by welding in the glut A to form the finished curve, which is cut by a hot set. The opposite or footstep end is formed from a mild steel plate, the angle being flanged over and cornered up square by welding on gluts of bar iron, 6 inches or 7 inches, at once. It is then welded to the main angle iron, care being taken to keep it in line with the whole length, and to trammelled centres already fixed. When finished, should it be too short, it is an easy matter to heat a short length and then stretch, but it is rather more difficult to jump it up accurately, these latter operations being of most rare occurrence. Fig. 251 is an example only of a box angle iron, the full lines having the flange outside, and the dotted inside. The corners A and B are finished first, that is, they are left perfectly square before the opposite end is commenced. The parallel sides are then cramped together, and the other two corners finished. The first corner must not be left until it is perfectly square, following this plan rigidly throughout for each corner. The inside flange is the easist to make, from the fact that there is only one half the scarfing and welding to perform.

Angle rings are seldom required, and those of heavy section and in numbers can be obtained weldless; but odd ones are made by bending the angle iron round segments upon the bending block and welding. To make an accurate ring is an accomplishment, the difficulty being the allowance or scarfing. Sometimes, instead of the usual rectangular or right-angled iron, acute or obtuse angles may be required, which are produced by drawing

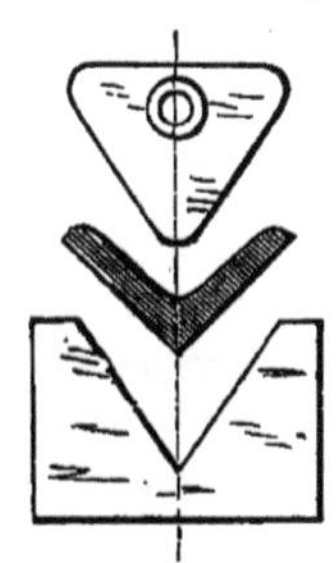

through blocks under the steam hammer as shown in the annexed figure; when of course the small quantity and short length required does not warrant the rolling of such a section.

One form of bending block is shown in Fig. 252; and upon it is placed the cab beading, which will at once bring before the reader's mind many applications to which it can be put.

General smithy.—The work of the general smithy, as far as the locomotive is concerned, is very much cut and dried, but in a railway shop, the locomotive work is supplemented by highly interesting jobs for other departments, which from a craftsman's point of view, unfortunately

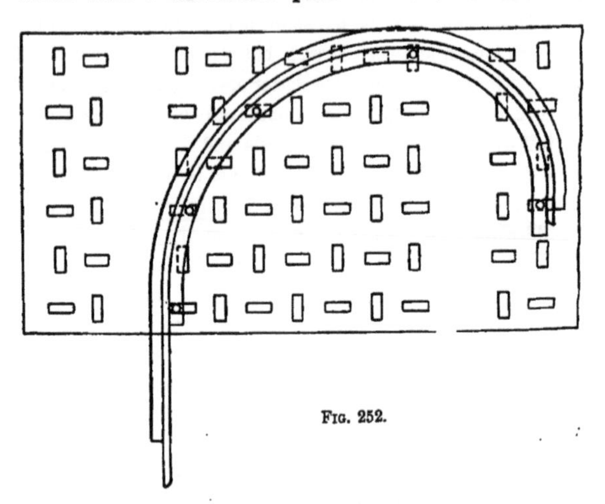

Fig. 252.

cannot be described here. In Fig. 253 is shown the lever at the end of the reversing shaft, which is received from the forge as shown at A and B. The first operation in the smithy is to fuller it with a round bar, as at B, then set down each side with a plate, cut off the surplus metal, and trim up on the anvil. Reheat and block it in the swage C, remove the fins, then trim and square up generally with fuller and flatter. Afterwards draw down the middle to width, thickness and length; then, after the little end is to section, cut off with a C gouge and set to the radius shown in the finished forging D, on the beak of the anvil. A well-known form of spring link is shown in Fig. 254. Two are made from one billet, 14 inches by 4 inches square, by hammering into it the triangular fuller, having the ⅝-inch square groove as shown at A. Afterwards the ends are drawn down, and a set put in for convenience in stamping in the blocks as at B. They are then straightened, trimmed up, and round fullers set in as at E, to the required depth, and then drawn down by plates to the finished section. The pin joint end is then roughed out in clapper swages, and finished

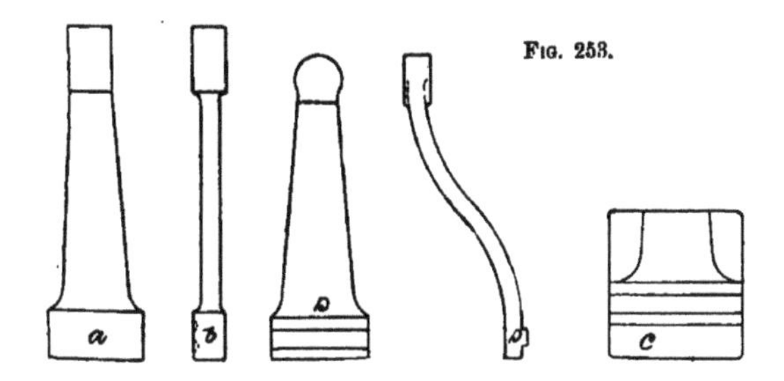

FIG. 253.

by stamping in a loose die, the fins removed, and finally it is cramped to the block, shown at D, resting upon the anvil, where it is generally trimmed and dressed up to template by fullers and flatters. The D link shown in Fig. 255 is another form of spring link. It is received from the forge as shown at A, the first operation in the smithy being to block it under the steam hammer in the swage B, with a round fuller on the top which prepares it for the finishing swage C, and it is then cut to length by the C gouge as at D.

In Fig. 256 is shown the punch and die for punching out of ⅞-inch plate one end of the safety-valve lever. G is a plan; H, the end elevation; K, a part section on A B of the plan G, showing $\frac{1}{8}$ inch lead on the punch at the centre, and L is a section on C D E F. There is a $\frac{1}{16}$-inch clearance in

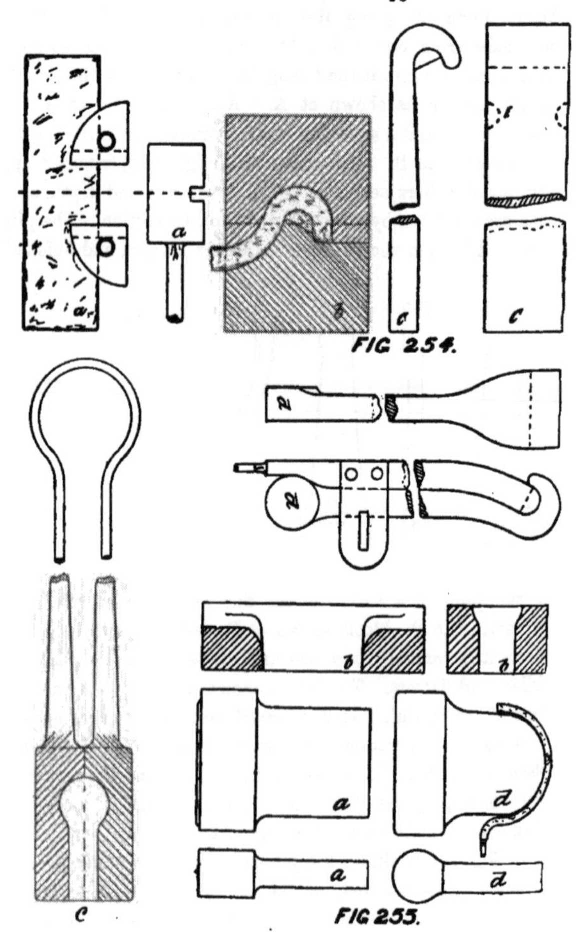

FIG. 254.

FIG. 255.

the die, and $\frac{1}{8}$-inch taper per side on the punch; and taking the figures as a whole, they are clearly indicative of their manipulation. After the punching has been dressed up, leaving, say, $\frac{1}{8}$ inch per side for milling, it is scarfed at M and the long arm welded on. The group of tools shown in Fig. 257 are for producing the screw couplings. C and C^1 are punched in a die, the form of which will readily present itself to the reader's mind. Two of these are required for

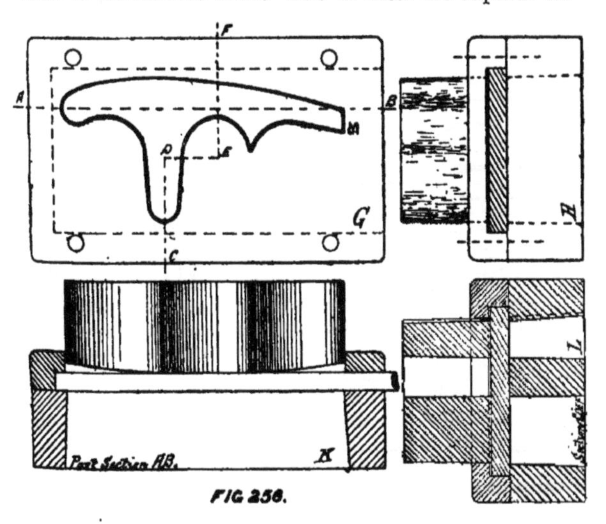

FIG. 258.

each shackle, and they are coupled direct by means of a bolt to the hook. D and D^1 show the punch and die for producing each end of the link H. It is made out of $3\frac{1}{2}$-inch by $\frac{3}{4}$-inch mild steel bar, and after each end has been stamped the middle is drawn down and rounded to $1\frac{1}{4}$ inch diameter. The screw E is made out of $2\frac{3}{8}$-inch round mild steel by first flattening under the hammer the short or shank end, then a round edged fuller is set in each side of the shank and each end drawn down; a few blows upon the flat of the shank and the whole finished in swages, one for the round section and another for the shank section. The nut G is formed as indicated, from round bar steel, say, about ¼ inch in diameter less than the finished product, because in this case the material will flow inwards to form the greatest diameter. Two pairs of swages are generally used, the first for roughing out, which

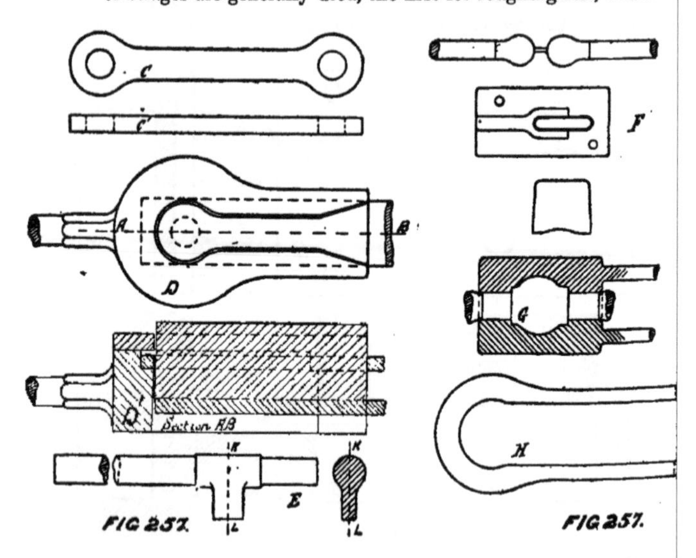

FIG 257. FIG 257.

is generally an old set worn by use, the second being the finished size. No machine work is required to be done to these nuts, excepting the drilling and tapping of the hole, because they are finished accurately, and with a good surface in the final die. The lever F, which is attached to the shank of the screw E, is shown as formed by swages out of 1½ inch square mild steel by 9 inches long, which makes two; but this joint, it may be stated, is now made by the drop hammer.

Two ends are formed simultaneously, divided and punched as indicated, the punch being shown separately beneath the die. Similar joints are a frequent job in a smithy, but of different dimensions and small quantity. In these cases the over-all dimensions are produced by fullers and flatters; then the bottom of the cleavage is formed by a hot punch, thus

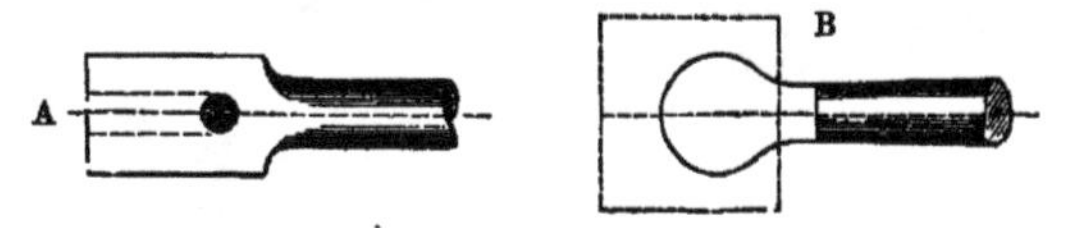

(A), and then cut out by a hot set, as indicated by the dotted lines. Afterwards a wedge is placed between for dressing and finishing up (B). See also valve spindle cross-head, Fig. 264, drop hammer. Fig. 258 may represent the ashpan and cylinder cock levers. Supposing the shaft is to be $1\frac{3}{8}$ inch diameter when finished, it is made out of $1\frac{5}{8}$ inch round, to

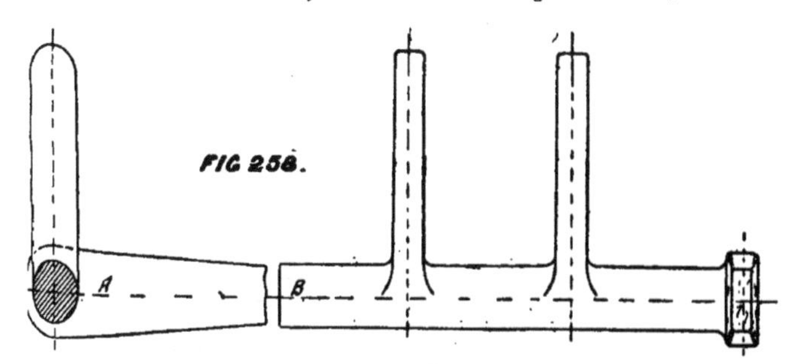

save jumping up. The long arm at the end is first welded on by setting back the end of the shaft with backing hammers, and hollowing out the arm, Fig. C, which is usually termed dabbing on. If it is a shaft of much heavier dimensions, these ends are fixed on differently in various shops, one instance being to fuller a hollow in the end, weld on the arm, Fig. D, and remove all surplus metal by the hot set. Another method is to fuller a hollow at the end, Fig. E, then draw down under the hammer and bend the arm over. The

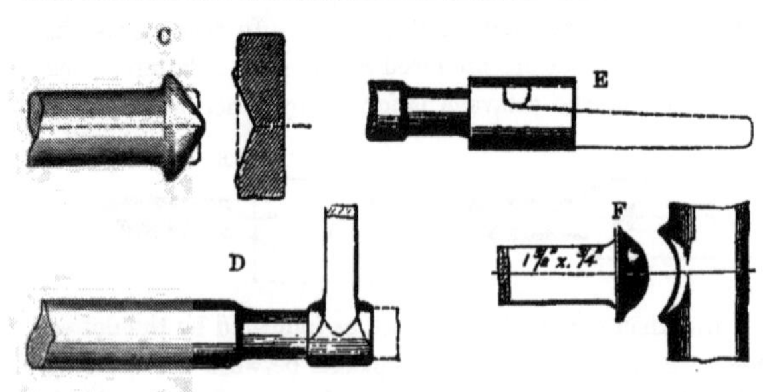

two short arms are then welded on by fullering out the shaft and scarfing the arm, Fig. F. After all the welding has been finished, the long and short arms are set to the required

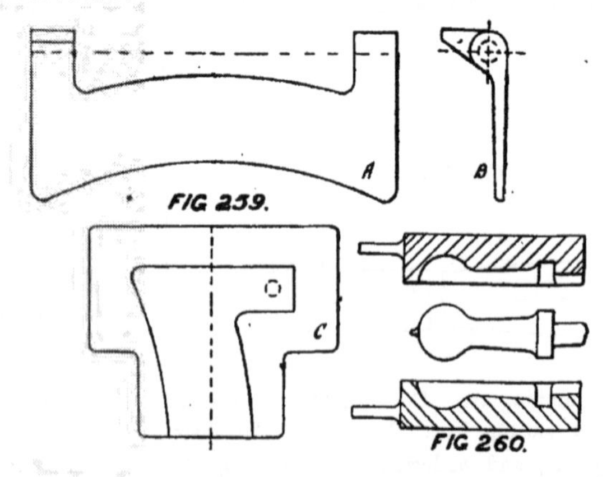

FIG 259.

FIG 260.

angle, then it is finished in round clapper swages, one pair having provision made to allow the short arm to protrude. The smoke-box hinges, Fig. 259, are made from 3½ by

$\frac{3}{4}$ by $19\frac{1}{2}$ inches long, drawn down by the aid of a taper plate to 4 inches wide and $\frac{5}{8}$ inch on the thick edge and $\frac{3}{8}$ inch on the thin edge, then set to a radius of 1 foot 7 inches, to which the lugs are welded. These are prepared by stamping, and then welded to the prepared plate in the block C. It is a well-finished job; the only machine work required is the drilling of the holes. The front and end elevation A B clearly indicates the whole arrangement, and also the use of the two different lugs. Fig. 260 shows the method of producing the boiler handrail stud, and Fig. 261 the handrail pillar. Each end of the latter is formed in its own particular

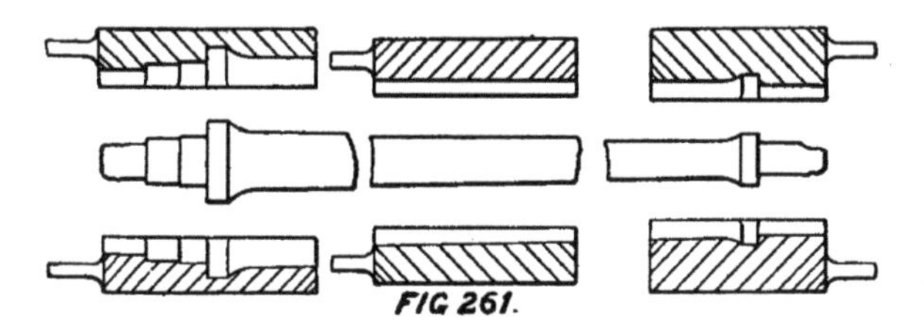
FIG. 261.

swage, and the intervening portion by a taper swage. The figures are clear, and do not call for any further comment.

In relation to stamping with the drop hammer, which has come in for a considerable increment of usefulness during late years, in many forms and capacities, it may be generally stated that all those remarks made upon swaging, in both loose and fixed blocks under the steam hammer, are equally applicable—Section III. Part I., Forge. Drop hammers may be obtained of various designs, and sometimes embrace many ingenious mechanical devices. They are sold by the makers of steam hammers, and also they are frequently rigged up by the intending user, out of convenient material in convenient places. When the latter plan is resorted to, it frequently happens that the well-known appliance in the tin shop is copied on a larger scale, and of heavier dimensions. In other designs the head is raised by means of an attached board, which passes through friction rollers, regulated by a foot-treadle, the upward progress being checked by adjustable collars on the upright. Sometimes arrangements are also provided to give repeated and variable blows. Fig. 262 gives four views of the hammer head, as an example, and Fig. 263, the top and bottom die. Mild steel castings make service-

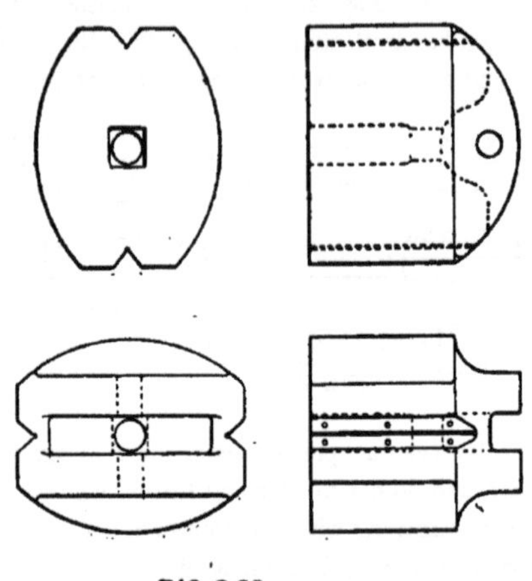

FIG 262

able dies, but of course whether the dies are forgings or castings, the object to be stamped has to be carved accurately in them. This is done by working rigidly to centre lines and machining as much as possible, planing, drilling or shaping; but the chief work lies in careful and patient fitting. After the fitter has finished they are braced together and tested, by running in a composition of lead and bismuth, and when accurately coincident they are finally marked, so as to facilitate the preliminary adjustment in the hammer.

When fixed for working, they are first tried upon a lump of stiff clay. It is reasonable to expect a production of a thousand articles without serious detriment to either die. Those shown have carved in them the end or joint of the

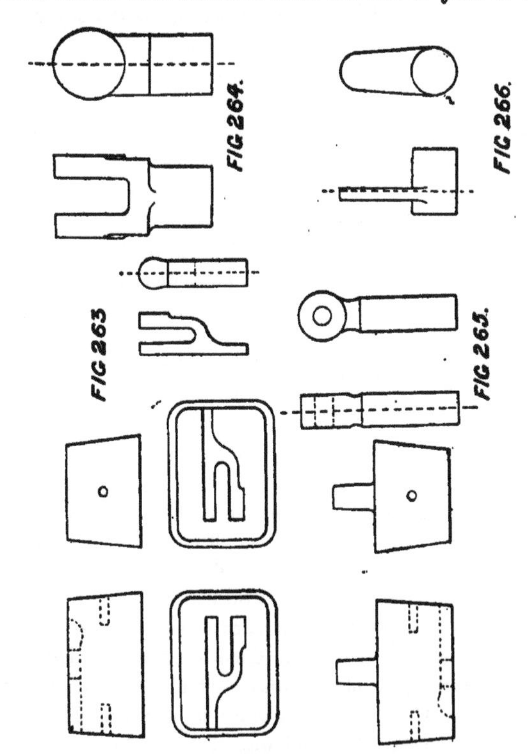

reversing shaft, the finished product being given in two views to the right. Other articles produced by this means are:—Fig. 264, the valve-spindle cross-head; Fig. 265, adjustable spring link; Fig. 266, a short-armed lever; Fig. 267, a small standard for communication cord, and also an infinite variety of signal work. The fin produced is removed by an auxiliary die, which is illustrated by Fig. 268, being the arrangement for the adjustable spring link, Fig. 265.

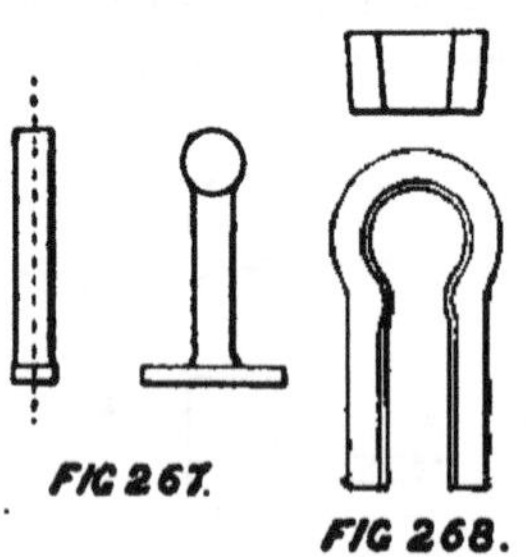

FIG 267. FIG 268.

Each engine is, of course, thoroughly equipped with all necessary tools, spanners from $\frac{5}{16}$ to $1\frac{7}{8}$ inch, box and gland keys, rakes, shovels, pricks, darts, chisels, &c. It is obvious that the cheapest method, taking all things into consideration, in the production of spanners, is the drop hammer, in which case they are made to the following section (G). The method now further illustrated, will certainly compete in economy of time, but there is the element of waste in the scrap formed by punching, and the spanner on the whole is heavier than that produced by the drop hammer. As there is no welding about them they are a strong job, and there is not any risk of the jaw breaking. The punching block and die for $\frac{7}{8}$ and 1 inch double-ended spanners are shown in Fig. 269, A being a plan and B the end elevation. They can be punched as quickly as the rolled or sheared strip of mild steel can be pushed through, there being two punches, one working whilst the other is drawn out and got ready to take its turn. After the whole order is punched, the $\frac{7}{8}$-inch ends are finished first, followed of course by the 1 inch. The finishing process occupies one heat only, the first operation being to round the edges in a pair of swages C, Fig. 269, immediately after it is put into the die D, a dummy placed upon the top to keep the die from spreading,

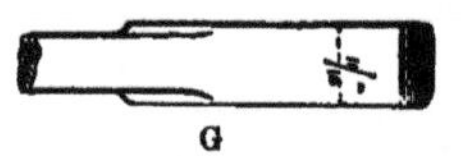

G

punched and finally dressed up on the anvil, of which, needless to say, little is necessary. Fig. 270 is a useful combination used in the production of chisels. It is one of the

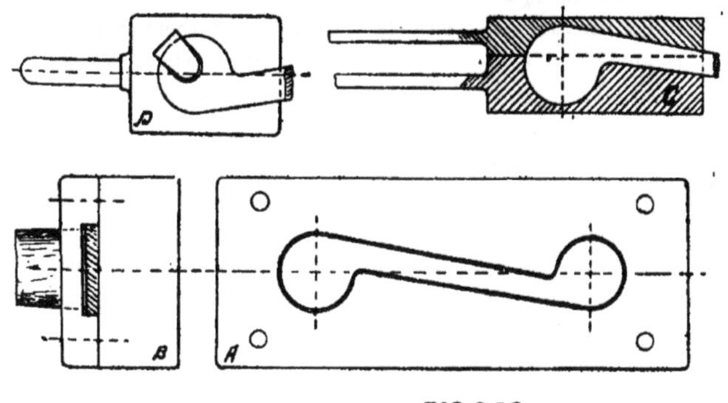

FIG 269.

many examples which might be illustrated to show how division of labour produces its own means to facilitate certain ends. It is used in heading, A being an adjustable guide for the length, B a peg which prevents undue wear and tear or injury to the swage C, which heads two at once. They are then sheared by a somewhat similar arrangement, the peg in this case preventing the blade of the shear coming upon the anvil, each combination tool weighing but a few pounds. After shearing, the chisels are drawn down upon a taper plate, and half-a-dozen can easily be made at one heat from bars of the required section.

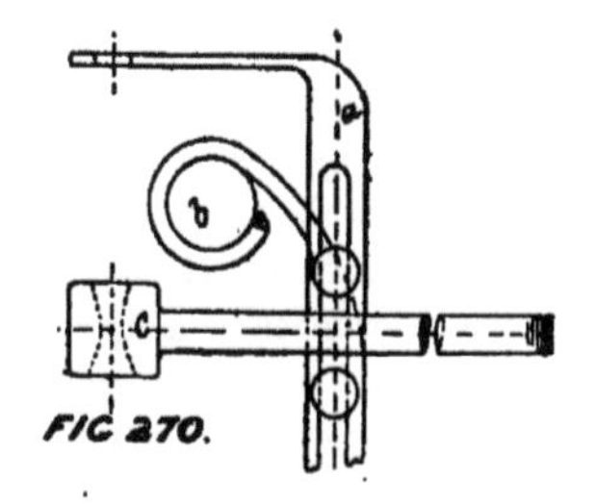

FIG 270.

The enormous quantity of nuts and bolts are made by

machinery, either by swaging or heading. The swaging machine has a range from $\frac{1}{2}$ to 2 inches by change of dies. Bolts with hexagonal heads are made from round bars, but those having square heads from square bars. The dies in the Ryder machine never close by $\frac{1}{8}$ inch, so that there shall not be any risk of damaging them. The machine is fixed in close proximity to the smith's hearth; immediately beyond is the hot saw, and in front of the latter is a square anvil, in which is placed the former A, Fig. H, for shaping the head, the chamfer being made by a snap. In the heading machine any length of bolt can be dealt with, but each has its range of work as to diameter, both for nuts and bolts, say $\frac{1}{2}$ to 1 inch, and $1\frac{1}{8}$ to $1\frac{5}{8}$ inch in bolts, the latter making from 1 to $1\frac{1}{2}$-inch

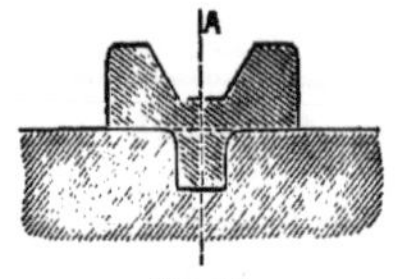

Fig. H.

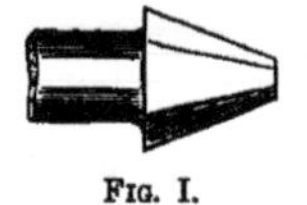
Fig. I.

nuts. Horsfall's heading machine first produces a "bunt," formed thus, Fig. I, which is then lowered down into the next pair of dies and finished by a snap and hammers, Fig. J. The nuts made in these machines are quite solid, and they do not make any waste. The punching from one nut goes to help to form the next, whereas in the Collier machine there are two punchings which go into the scrap box. The range of the latter machine is up to $1\frac{1}{4}$-inch nuts and washers. In this case bar iron must be used about $\frac{1}{8}$ inch thicker and $\frac{1}{16}$ inch narrower to allow for compression and flow during the act of punching. Rivets are made in machines of the De Bergue type, which are heading. The thickness of the nuts is the same, whether hexagonal or square, and generally equals the diameter of the bolt, excepting in some cases, such as gland nuts, where it is a practice to make them $\frac{1}{8}$ inch thicker. The thickness of lock nuts is two-thirds the diameter of the bolt, omitting fractions below $\frac{1}{16}$ inch, which is also

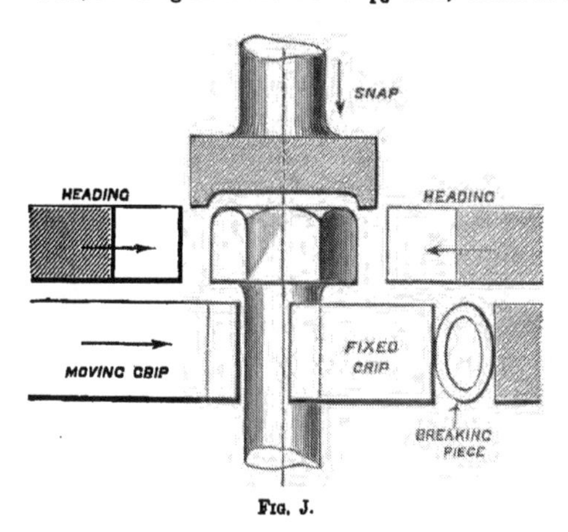

FIG. J.

the case with snap heads, the width of the latter being equal to that of the nut across the sides, Fig. K. The large radius A equals twice the diameter of the bolt, and the small radius B half of the diameter. The angle of countersunk bolts is made 30° with the vertical, the diameter of the head being equal to the width of the nut across the sides; therefore the depth determines itself. The nuts are to Whitworth's standards. The ferrules of the tubes are made from steel strips

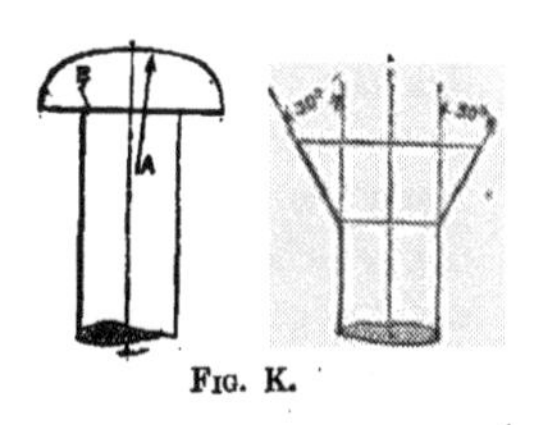

FIG. K.

rolled to section, cut to length by the smith and welded by an oliver, on a mandril in a die. This method has been surpassed by cutting steel tubes of required diameter to length, and putting the necessary taper on by a drifting machine.

Spring smithy.—The work of the spring smith has been considerably curtailed by the introduction of a compact machine, which combines all the processes of shearing, tapering, spearing, or shaping the ends of the plates, punching, slotting and nibbing, for which, up till now, if machinery has been employed, it has been requisite to use separate machines. It occupies a space of about 12 by 8 feet over all, including arms for resting plates upon and adjustable guides for lengths. It consists of an upright frame carrying three slides in the centre, working respectively to a couple of dies for spearing or shaping the ends of the plates, putting in the nibs, and a punch for cutting the horizontal slot. A fourth slide is utilised for the working of a circular punch to form the centre indent for the $\frac{3}{4}$-inch set screw, and above this is a pair of shears for cutting the bars to length, to which is attached an arm for an adjustable guide for those lengths. At the opposite end of the machine is a pair of rolls, one of which turns eccentrically and tapers the ends of those plates which are required for certain classes of springs. Alongside this, fixed to a back centre, there is an arm which rests upon a cam, and its duty is to close in again the metal which has spread during the thinning process. If only one width of plate was used, this last piece of mechanism would not be required, as a groove in the roll would prevent the metal from spreading. The whole combination can either be driven from the shafting, or, where this is absent, by a small vertical engine with ordinary gearing, and is operated by one man. For all ordinary grades, say up to about 48 tons per square inch tenacity, this machine operates upon the plates at ordinary temperatures, but should a harder grade be required, the plates are got to a dull red heat before being machined, and then portioned out in complete sets to the various smiths. The pin ends of the spring back are stamped in blocks from a good quality iron, which in the long run is the most economical, because they can be cut out of damaged backs and welded on to new ones, and should the holes wear they can easily be closed again, whereas if made from any grade of steel beyond a mild, it would be more difficult to cut them out and close the holes, or re-weld in new backs. Solid ends are prepared by turning the plate round a piece of ½-inch iron, welding up and finishing in a swage fixed on the anvil Fig. M. Each smith has a setting table about 5 by 2 feet 6 inches by 4 inches thick, and a block to sketch, Fig. L, also a good set of cramps to assist in placing on the buckles,

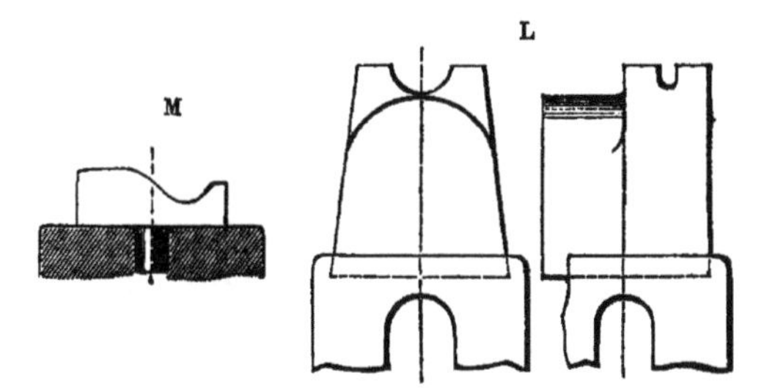

and one oil tank 4 feet in diameter by 5 feet 6 inches deep serves for two hearths. The use of the block chiefly lies in rectifying the cambre of plates after tempering, for although the tendency is to straighten, it is never known in which direction the plate may go or behave during slacking—one edge may go more than another—also for bringing to cambre any plates that have been damaged through accident or unusual wear and tear. Each smith makes his own backs by welding on the ends, which have been drilled for the pin, and clearance slotted for the link. This job alone shows the smart action required in welding steel of moderately high carbon, for the backs are a special welding grade not quite so

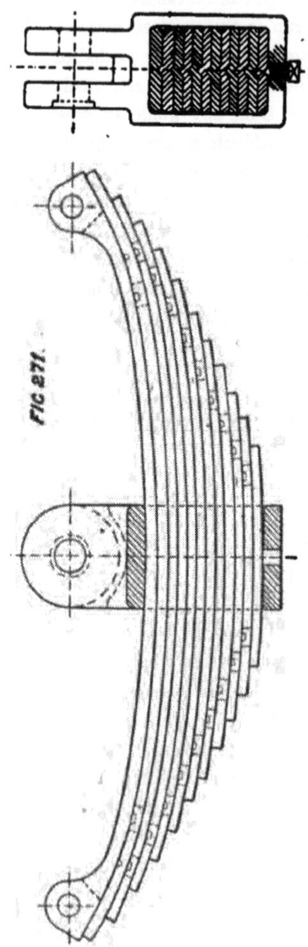

hard as the body of the spring. After each end has been welded on, a composition being used to aid this, the whole is got to a uniform heat and the cambre set in to a template, and then followed on by each succeeding plate to the preceding one. Each, as it passes through the smith's hand, is hardened and tempered in oil, the latter operation generally depending upon the behaviour of the heated plate when rubbed with a piece of wood. Judgment in this case depends entirely upon the knowledge of the grade of steel, any change in which can easily be detected by its ring when struck, but sometimes in the case of safety valve springs, one or two are tried in the testing machine. The collection of plates is then cramped together, and the buckle, Fig. 246, p. 145, is placed on hot, compressed upon all sides simultaneously in a specially designed hydraulic press, and then quenched.

Every spring is tested to a pressure sufficient to take out the cambre, not more than $\frac{1}{8}$ or $\frac{3}{16}$ inch permanent set being admitted. The finished spring is shown in Fig. 271. The safety valve spring consists of six coils of $\frac{25}{32}$ inch diameter steel bar, one half of a coil at each end being turned inwards for suspension. The length between the points of suspension is equal for 140 lb. and 160 lb. per square inch, the difference in the tension being effected generally by temper, but it may also be regulated by composition of the metal or reduction in sectional area. It, and the mandril upon which it is made

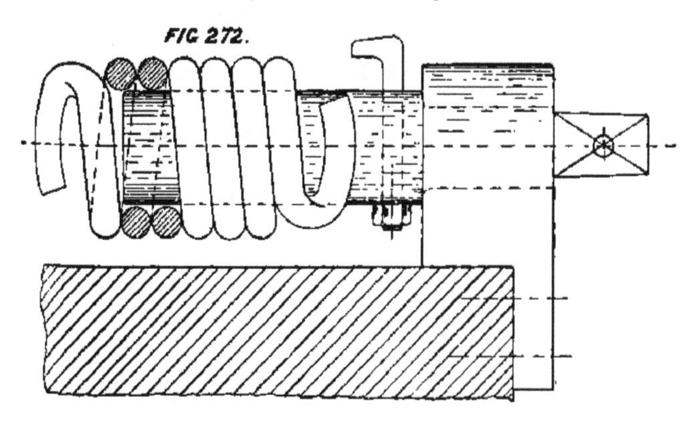

are shown in Fig. 272, which will also serve as an example of the production of this class of spring. Owing to the strong section of some volute springs, the motion of the mandril must be produced by mechanical effort.

Among the processes accessory to smiths' work, it may be stated that tempering in its broadest sense embraces hardening, case-hardening, annealing and toughening. Considerations to be regarded when effecting a certain degree of temper are resistance to abrasion, the sustaining of a great pressure upon a cutting edge, and also the elasticity of the remaining portion of the tool or article in hand. The intrinsic value and total efficiency of a tool depends much more upon the manipulation of the operator during its production, and its tempering qualities, than upon its direct hardening property. The former is adjusted to suit the nature of the work it has to perform, and is really the obtaining of a hardness with resilience, less than the ultimate or dead hardness. The heating should be as uniform as possible, otherwise there will be an unequal expansion, and sometimes to such an extent as to render the finished article of very little value. The rapidity with which the heat should be developed depends entirely upon the size and the utility of the article; but in most cases it is safe to state that slow and even heating should be resorted to, with as little decarbonisation going on as possible, but if the article is of large dimensions and even section, it may be of considerable service to heat the exterior rapidly and evenly, then, after cooling, the interior will be of a lower temper than the exterior, and consequently the article will be backed up with an interior of greater resilience than the exterior. Hardness is the direct result of quenching, which should be carefully done, so as to prevent as much as possible, warping and cracking; rectifying any deviation from the straight being possible by careful insertion in the cooling medium. Water should be kept clean, and it may be heated to about 100° F., 38° C., with a diminishing risk to cracking. Dip straight articles vertically, and here two opposite influences occur, for if the article is allowed to remain still, the spheroidal state follows—if the author may so put it—and consequently the plate is not so hard, whereas if moved about there is an unequal hardening produced, that portion leading being the hardest; hence, it has been found most beneficial to inject a full stream of water upon the article in specially constructed tanks. Undoubtedly, water that has been used for a considerable period has a superiority for hardening, which may be chiefly attributed to the fact that with use the dissolved air has been expelled. The ultimate hardness depends, first, upon the amount of combined or hardening carbon which may be taken as a true alloy of iron, and as the content of this is higher or lower—tenths per cent.—so is the degree of susceptibility to hardness; secondly, upon the difference between the temperature of the article and the cooling medium, and also the conductivity of this medium. The existence of the carbon in different forms is a much discussed point, for by the action of solvents more than one modification is discovered, two of which only it is proposed to deal with, as sufficient for our purpose.

The effect of annealing appears to be the reduction of the combined carbon, Section II., Part II. p. 78—a separation out in the interior, and a partial disappearance upon the exterior. Sudden cooling acts exactly opposite, retaining it in its combined form. Compression appears to bear an important influence, for as the content of carbon increases, the elastic limit is also correspondingly increased, and therefore a more powerful and rigid compression takes place; that is, the outer layers, by virtue of their elastic limit, refuse to surrender to the resistance to compression offered by the inner layers, and therefore the greater the differences between the temperature of the article and the cooling medium, the more serious is the trial to the elastic limit; and consequently, in steels whose chemical composition has raised the elastic limit to be their ultimate strength, it is only natural to expect serious loss by cracking. Finally, hardness and brittleness go together, according to chemical composition, and tempering restores to a certain extent the elasticity, without destroying to a serious degree the hardness. The hardening is generally confined to the immediate vicinity of the edge, and as the heat travels forward the well-known colours develop, indicating certain temperatures, and as the desired colour appears, its progress is finally stopped by total quenching. Although this colour may indicate quite definitely the temperature of the material in hand, it cannot possibly indicate the exact temper, because this depends upon the chemical composition, which condition must be known to the operator, or experience gained upon this point. The foregoing has been stated upon the assumption that hardening upon sudden cooling is due to a change or difference in the condition of the carbon, but the most recent theories propounded are, that hardening is due to an allotropic change in the condition of the iron itself—in fact, that iron is a polymorphus element like sulphur; but whether this phenomena is due to chemical reaction or allotropic changes, or both, it is not for the author to discuss in the present work.

Case-hardening is a short period cementation process, the theory of which is well known and unnecessary for discussion here, the benefit being a steely surface with a soft tenacious backing. It is produced by packing the articles to be hardened in an air-tight box with animal carbon and cyanogen compounds, then gradually raising the temperature and maintaining it at a red heat until the required depth and degree of hardness is obtained. The crank-pins, and the end of the coupling-rods that require to be case-hardened, generally receive about twelve and eighteen hours' heating respectively; the motion and pins seldom require more than twelve hours, and in certain classes of steel for milling cutters about four to six hours, the cutters being quenched in oil. A small test sample is placed in each box, about ¾ inch diameter and 5 inches long, and each sample is registered, together with the articles case-hardened in the same box, which is heated in a small Siemens or other suitable furnace. Soda and lime or other alkaline substances assist the process, which in works other than locomotive, especially in relation to cotton machinery, is a large undertaking in itself, as much as 6 tons of bone alone being used in one week. A rapid production of this hardness may be obtained, but only to a slight degree in depth, by rubbing the heated surface with prussiate of potash and quenching. Annealing has been thoroughly dealt with in relation to steel castings, page 72, and toughening bears a direct relationship to annealing; but toughening by water quenching has been made the special study of a material with a high percentage content of manganese by Mr. R. A. Hadfield, of Sheffield.

SECTION IV.

COPPERSMITHS' WORK.

THE solid drawn seamless tube has entirely dispensed with the old brazed seam for locomotive steam pipes. They range from $\frac{3}{4}$ inch to 7 inches or 8 inches external diameter, and are a great advance towards the perfect steam pipe, but, owing to the almost universal interest in the brazed seam, it will not be omitted from this section. The seamless tube must be of uniform circumferential thickness and perfectly round. A piece 30 inches long, annealed, and then filled with resin, must withstand bending until extremities meet without showing defects; but if the sample is not annealed it must sustain a deflection of 3 inches when placed upon supports 20 inches apart and loaded in the centre, without showing local contraction, cracks or other defects. The circumferential strength in tension of a hard drawn seamless tube may be taken at about 30 per cent. above that of a carefully made one with a brazed seam, and, even after annealing, this tenacity is only reduced by about 15 per cent., which is due no doubt to the extra mechanical work imparted to the material by drawing. Boiler tubes successfully withstand an internal pressure of 800 lbs. per square inch, and an external of 250 lbs. per square inch, without leakage, being $1\frac{3}{4}$ inch external diameter, and having a cross-section of about ·52 square inch.

Owing to the fact that copper rapidly loses its tenacity at high temperatures, and that the range between a brazing heat and that when it becomes red-short, or rotten, is narrow, it follows that the value of a pipe with brazed seam is in ratio to the skill of the workman, and perhaps no operation depends so much upon the operator. If this brittle heat should be produced to only a small amount, then any degree of internal stress would produce cracks of various grades of importance, most likely not sufficiently developed to be detected by the hydraulic test, but which would grow by wear and tear, vibration, repeated expansion and contraction, until finally, if not discovered in time, the pipe would burst. The next matter of importance is the scarfing, the length of which depends upon the size and thickness of the plates. In any case both the edges should be bevelled, which, if of medium length, will always make the best joint, because long bevels have a greater tendency to burn. The machine planed scarf makes the most perfect joint, although the paening hammer is responsible for most jobs. It is important to bring both scarfs together so that no gutter is formed; and also it is imperative to bevel both edges, although many pipes may have been made with the inside edge left square, in which case the tendency is for the sharp square edge to cut into the material overlapping it, and increase the liability to rend. For brazing a good clear fire is required, and if there has been any tinning going on before, the operator is careful to remove all traces of that metal from the fire and its precincts, for should he have the misfortune to get a little scrap of tin on to the joint, it will percolate its way through the copper, or combine with the spelter, burrowing, "rat-like," and spoiling the job. This can be easily demonstrated by placing a small bead of tin upon a thin copper plate over a fire, when it will rapidly eat its way through—a fact readily explained, and realised after a perusal of Part III. of Section II., "The Brass Foundry," page 96; and needless to state that in a locomotive shop where the work is well regulated it is a most rare occurence. Clean metallic surfaces are indispensable, because solder spelter will not adhere to anything dirty or greasy, but with tinned surfaces no particular care is needed, as the solder will adhere readily and quickly. There are many hard solders in use adapted for special jobs, either copper, iron, or steel; that for the latter may, of course, be stronger than for the former, but in any case it should be tough and fusible, a lower melting point than the materials brazed being of a necessity, for then risks of burning the scarfs are reduced to a minimum. Borax is generally the fluxing agent used, then, if all due precautions have been observed, the brazed seam of a copper pipe is as strong as the rest of the material; but the extension measured on given lengths will be, even in the best cases, 50 per cent. less than that upon the rest of the material, excluding the seam. This extension and contraction of area takes place upon each side of the seam, leaving it almost perfectly at its original area, breaking in the best joints parallel to the seam, in a straight line at the junction of the bevel with the original thickness, but in the inferior joints the rend is very jagged, and it is to be generally observed that this jagged rend is due to either hollow places where the solder has not penetrated, or where either dirt or flux has been entrapped, and therefore for these reasons 15 per cent. should be allowed for reduction of tenacity. This of course, relates to copper, it being needless to take this point into consideration in brazing mild steel sheets, as no pressure is required behind them. If that were necessary, welding would enter into the question.

The internal pipes consist of solid drawn tubes $\frac{3}{4}$ inch to $2\frac{3}{8}$ inches diameter, and of Nos. 10 and 11 I.W.G. for the injector and ejector steam, exhaust and delivery, and the blower or jet pipes. A set is put in the ejector exhaust pipe, to escape the wash-out cocks on the side of the fire-box.

The injector and ejector steam pipes obtain their steam from the top of the dome, by the aid of a brass elbow instead of a bend. The injector delivery pipe conveys the water to an extreme distance from the greatest concentration of heat and sufficiently far from the dome, thereby minimising a tendency for the presence of watery vapour in the steam. The blower or jet pipes receive their steam from the safety valve seating, and deliver into the chimney. The work done on these pipes is comparatively little, and that to wire templates, which are made to a full-size drawing produced upon the copper-shop drawing board, this also being the method of working for all bending or sets on pipes, whatever their size. They are softened in the locality of the work, and bent by hand, with a bending fork, clip and bar, through an eye lined with lead, fixed to a stump. In most cases, after having been softened for bending, they are stiffened up again. The external pipes for the sand gear, injector, brake, or water-pick-up are treated in a similar manner to template; but straight pipes are simply cut to length, and their couplings brazed on ready for the engine. The internal main steam pipe, Fig. 273, is a straight piece 4 feet 9 inches by 5 inches by 7 I.W.G., to which the cone is bored a good fit and fixed to one end, then brazed in an upright position in the pot fire. The cone itself, Fig. 274, is covered upon the outside and part up inside with a mixture of fire-clay and plumbago, a hollow recess being formed at the top, which is charged with solder and then flushed; after which it is dressed and bolted up in its place in the boiler, the flange, Fig. 274, fitted on at the tube plate, then brazed and tested ready for final fixing. The smoke-box steam pipe, Fig. 275, is also 5 inches by 7 I.W.G., softened, filled with resin, lead

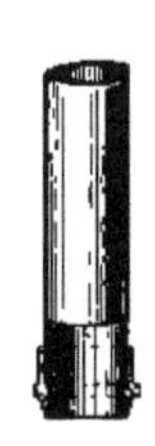

FIG. 273

being used in urgent and light jobs, then bent to templates in a suitable hydraulic press, round a mandril of the required

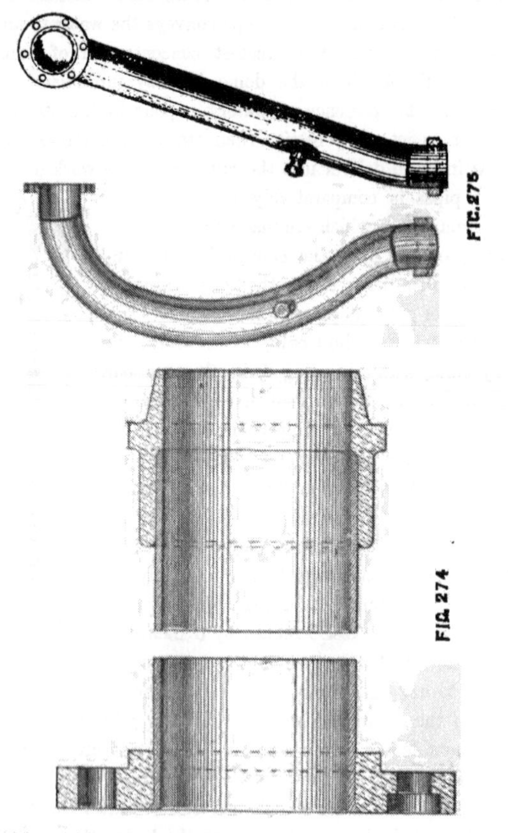

radius. Holes are then drilled in for the lubricator pipe, the saddle of which is well bedded on by fitting, cramped by a piece of steel wire binding, or the hole in the pipe is tapped, and then the saddle is held by a set screw, when precautions must be taken not to allow the spelter to strike the threads. After the saddle and flanges have been brazed, the pipe is then fitted up in the model of the standard smoke-box, which is kept in the copper shop, so that when the pipe goes to the erecting shop it is at once fixed in its place.

All pipes are carefully tested to 200 lbs. per square inch for locomotive work; but for lower pressures double the working pressure is the usual test. The brazed joints, flanges, or cones are hammered slightly whilst under pressure, which facilitates the discovery of any flaw or discrepancy.

The flanges are made stout and bored parallel, slightly countersunk at one side to give the spelter a good chance to strike or flush in, thus giving a good bearing for the pipe. In order to save trouble and expense of flanges, sometimes the socket joint is used, the only example of which to be found in locomotive work is when that portion of the pipe just above the flange, on the steam chest end of Fig. 275, has worn thin, then it is removed and a new piece put on by aid of this joint; or it may be some lubricating pipes are jointed up this way. It is a good joint, but should not form a portion of a bend, or even be in the immediate neighbourhood, and great attention should be bestowed upon the solder flushing to the bottom of the socket. Cases are known in which copper or other wire is wound round the pipe, or hoops placed at given distances or wound at a given pitch, to minimise the risk of bursting and—in the event of such occurring—to prevent as far as possible the extension of the rent, this being more especially applicable to the pipe with a brazed seam.

The principal accessory operation in this department is the preparation of the dome and safety valve seating covers, also the mouldings for the fire-box and smoke-box corners, all of which are made from the best mild steel sheets, 14 I.W.G. thick. In some cases the dome covers are made in two pieces, blocked and brazed, but necessitating the use of a heavier material. In another, the one about to be described, they are made in three portions, the following sketches representing the progress of their formation, templates being first made, all sheets marked out by their aid and sheared to shape.

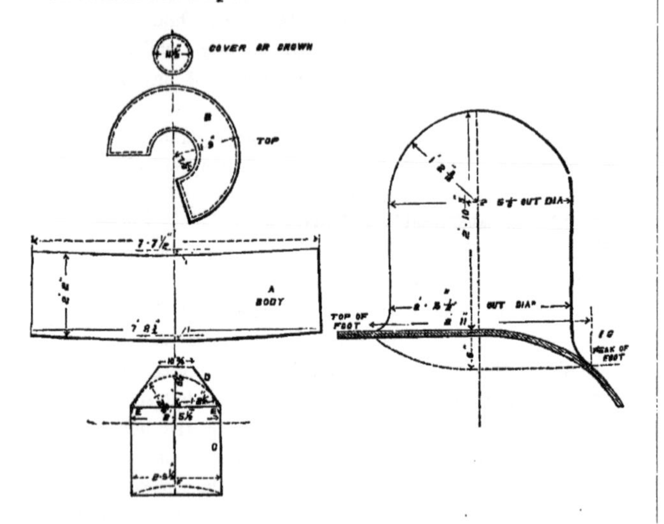

The two portions A and B, forming the body and top, are then thinned upon their edges, which form the joint or seam, with a paening hammer on an anvil, and afterwards annealed. These two are then passed through the rolls, which brings the paened edges or scarfs together, which are then "cramped" or dovetailed, each being about ⅝ inch by about ¾ inch wide, and then brazed, the spelter for this job being fusible and tough. The formation is then C and D.

The body is then worked in at E by two or three courses at the pneumatic hammer and the top of the foot cut as indicated by the dotted line F. The foot is then worked out at the pneumatic hammer by gradual courses to templates and radius plate G. The top D and the crown are then formed in a hollowing block, which draws in the extreme corners, then thinned on a suitable stake, afterwards cramped and brazed. The whole is then planished to template where required by hand hammers on suitable stakes or dolleys, ready for the erector.

The production of the safety-valve covers is a repetition of similar operations, the following sketches clearly indi-

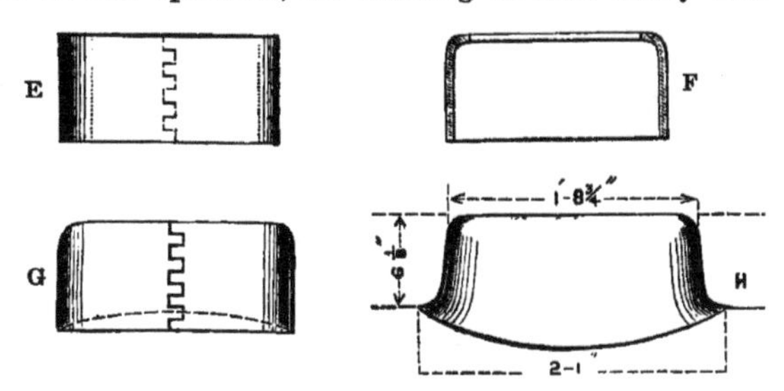

cating the progress, step by step, of the work, the final operation to this job being to cut out the clearance space for the valve pillars. The body is formed by one rectangular piece, 5 feet 5¼ inches long by 10 inches deep, and the top 1 foot 6½ inches diameter. E, body thinned, cramped and brazed; F, body worked in at top for about 1½ inch by pneumatic hammer; G, body with the top brazed on, dotted line showing removal of the top of foot; H, final, using radius plate as for dome, by gradual courses at the pneumatic hammer.

The corner moulding for the fire-box is cut to the template and then brought to shape by hand hammering in a suitable hollow block, then planished all over and fitted in its place on the engine, which consists of making it a neat job to the various boiler mountings which happen to come in its vicinity, Fig. K.

Those for the junction of the boiler barrel with the smoke-box are usually produced by a set of rolls; but in the

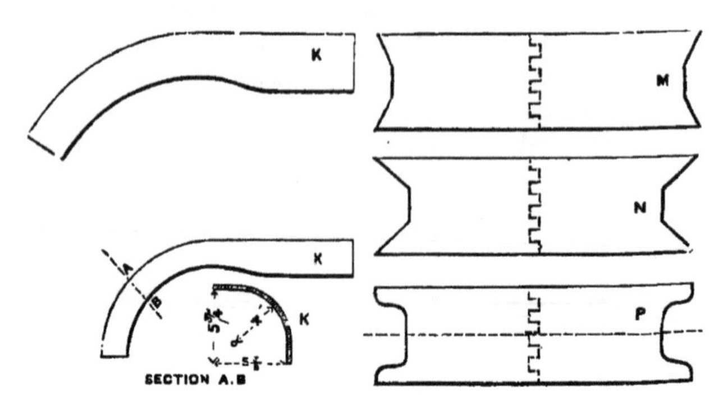

event of their absence, for whatever reason, the following would be the best and cheapest method of production:—A sheet should be selected broad enough to form two pairs, rolled and brazed to form a ring. It is then taken to the pneumatic hammer and a course worked out as in Fig. M, then annealed, and afterwards a second course worked up —Fig. N. It is again annealed, and roughed out at the pneumatic hammer to Fig. P, after which it is placed on a suitable stake and finished by hand planishing to the exact radii, Fig. R, cut round the middle, and the two pairs are ready for fitting in their places.

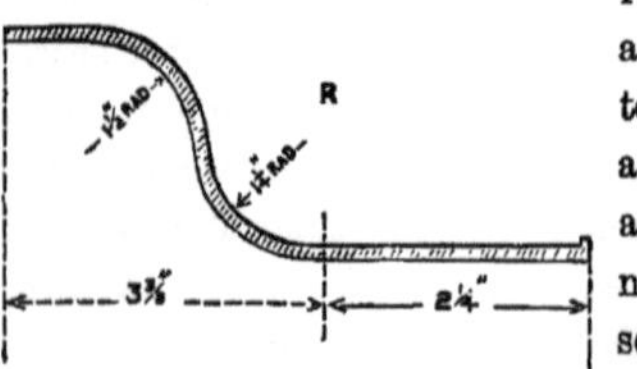

SECTION V.

THE MACHINE SHOP.

MACHINE shop economy is summed up by developing every machine to its utmost efficiency, and obtaining the maximum output from each, consistent with the requirements and nature of the work, which can only be achieved by a good system of tool work. A fenced portion of the shop should be devoted to the manufacture and storing of tools for the whole of the shop requirements, discriminating between poverty and excess, and discountenancing all waste. An attendant should be appointed to supply the wants of the men, who come provided with the necessary check from the foreman, with the various stores in his charge, all of which are booked to the recipient and again when returned, explanation having been rendered to the foreman for any detrimental effect the tool may have sustained, which is intimated by another check to the storekeeper. Taps, stocks, dies, templates, gauges, twist drills, mandrils, milling cutters, &c., are all dealt with in this manner, suitable headings being fixed in a book for the purpose, together with spaces for the workman's name, number, dates, and remarks as to condition.

What is a shop without system and organisation but simply chaos? whereas with them everything is reduced to order, every man knows what to do, when and where to do it. System does not require any more workmen, but it reduces the work of existing hands, and everything is done well, because each is individually responsible for the particular operations in his care; but, at the same time, this responsibility is reduced to a minimum, because every individual knows the system, and that it will be rigidly carried out. In fact, it is impossible to grapple with large concerns without it, and even small ones are, or soon would be, utter failures. It also insures that nothing is interrupted, not only in one shop, but every portion of the works, by the absence, for whatever reason, of foreman, man or apprentice; also every machine is fixed to plan, so that heavy or light work can be carried on most expeditiously, and suitable crane power or lifting tackle is found exactly where it is required.

The beneficial effect of accuracy in tool work is well and easily illustrated by a twist drill. The circumferential speed for $\frac{1}{2}$ inch to $\frac{7}{8}$ inch diameter is 20 to 30 feet per minute *on* mild steel, and a good feed is about $\frac{1}{100}$ inch for each revolution, that is half that amount per lip for each revolution; consequently if the drill is ground with uneven lips, the whole cut comes upon one edge, and therefore shortly it is damaged, and the driller reduces the feed of downward traverse until this one edge cuts well, which is apparently, to one-half the feed; and therefore even to drill at the smallest cost, absolute accuracy is required, concentric and true throughout. Each edge must be of equal length for obvious reasons, and have the same angle with the centre of the drill. Grinding lines are now dispensed with, owing to the use of mechanical grinders. The angle of clearance must be studied with a view to avoid rubbing and digging; and a test—the only one that the author is aware of—for the accurate grinding of a twist drill, is that it shall refuse to pass through the hole it has drilled by virtue of its own weight. The speed for twist drills may be taken at about 25 feet per minute, the feed being at the rate of 100 to 120 revolutions per inch of downward traverse, and say, 100 to 120 $\frac{1}{2}$ inch holes 3 inches deep may be drilled through wrought iron or mild steel; but taking the whole shop through, only half this quantity may be reckoned upon before the drill requires to be re-ground. Rose bits are not entirely dispensed with, as it is found useful to keep a few for the pin holes in the motion. Two very important points respecting all tools are the cutting and clearance angles; but much has been written elsewhere —in fact, whole books of no mean dimensions—upon this edge question, including all tools with one or more cutting-edges. It may be stated that in all ordinary lathe work deep cuts and coarse feeds are first principles, one roughing and one finishing, bringing down the speed to suit the cut, rather than suit the cut to the speed, for then the greatest amount of work will be done in a given time; that is, let it be a maximum of feed rather than speed, and have the finishing a good sliding cut of greater feed than the roughing. The old *régime* is well described in 'Extracts from Chordal's Letters' as "the wiry, rough tapering and changing surface of the fine feed is the standard for rough work, and these defects corrected by the file are the standard for the nicer work."

Any lengthy dissertation upon screw-cutting, gaps or stays in lathe work, does not come within the limits of this section, in fact it is not required; and the same remarks relating to the cutting edges are equally applicable. It is impossible to go through and illustrate everything in the machine shop, because of multiplicity, time and space, also it is quite unnecessary. Generally, descriptions of machines that do the work will be conspicuous only by their absence, because we are dealing with methods and suitable illustrations of advancing work most expeditiously, at a minimum cost, on those machines well known to all constant readers of *The Engineer*, and its well illustrated advertisements.

Milling machines now take pre-eminence of all tools in

any shop, and the *raison d'être* for the efficiency and economy of the milling cutter is now so well understood, that it would be utter waste of time on the part of the author and of his readers to enter into a lengthy argument upon that point. Too much cannot be said, and it is almost impossible to push the utility of the universal machine too far, and the adaptation of special machines during the past ten years has been quite phenomenal. One of the first essentials is to well man the machines, and then keep the cutters in order, for success depends entirely upon the facility for production of, and re-grinding the cutters, it being an absurdity to use a cutter beyond its profitable period of service. A 3-inch cutter will do probably, on an average, from sixteen to eighteen hours' work before re-grinding, which only then takes a few minutes, because, in most cases, not more than $\frac{1}{100}$-*inch* removed will restore the cutting edge, providing the cutter has been fairly used; but a 10-inch cutter may run quite 100 hours, the cost for re-grinding in each case not exceeding sixpence. It is also an axiom that a repetition of work is a necessity; consequently special machines are designed with the cutter spindles placed in various positions to suit that work. Also the emery grinder has reduced the cost of the production of cutters to a reasonable and justifiable limit.

For general work the plain cylindrical cutter is the most effective and least expensive. As compared with a facing cutter, there are teeth upon the periphery only; consequently it is less costly to make and keep in repair, also there are not any corners to chip or burr up by excessive stress of work, and facing cutters for gauge work are useless after re-grinding unless the patent adjustable split cutters are used. The ratio of the cost of labour brought to bear upon the material in producing a first-rate tool, does not in many cases admit of a comparison with cost of material; consequently nothing but the best quality should be employed, which is frequently obtained only by experimental data upon the work to be done. Experimental attention upon this point and also upon suitable angles for the teeth, both cutting and clearance, is bound to repay the operator to an almost unlimited extent. It appears that up to the present no acknowledged rule has been adopted regarding the pitch of teeth, which now simply rests upon the individual experience of the manager. For all practical purposes, whether for periphery or face, a pitch of about $\frac{3}{8}$ inch or $\frac{1}{2}$ inch will give satisfactory results with angles cut to 45° and 12°, but re-cut to 40° and 12°, with an angle of clearance or relief of about 3° to 5°, varying for different materials, and this angle may be taken about the same as that for a turning or other machine tool. The angle of relief is maintained when re-grinding, by placing the centre of the emery wheel either a little below the centre of the cutter if upon one side, or a little to the left of the centre line if the emery wheel is placed above the cutter. A spiral groove or thread of about $\frac{5}{8}$-inch or $\frac{3}{4}$-inch pitch, forming what is generally called a "lob" cutter, will enable it to do rougher work by relieving the teeth of the shavings in detachments; but a continuous tooth is preferable for finishing, the angularity of the spiral of which ranges from 1 in 30 to 1 in 60 inches, the finer pitch being the easiest upon the machine, its cut being more continuous and less intermittent, so that each tooth does its work gradually, and consequently there is less wear and tear upon the machine. The direction of the spiral is arranged either right or left hand, to force the work upon the table and the spindle against its bearing, the direction of rotation deciding this point. It is beneficial to use cutters with the least possible diameter consistent with the work they have to perform, because, naturally, they are less expensive to manufacture and maintain, also heavier cuts can be worked with less strain upon the machine, owing to the smaller leverage due to less radius. Again, as shank cutters and "chambered" ones are frequently interchangeable with two or more machines of their kind, in order to get standard spindles, it is possible to get a set of cutters on a machine for which they have not been designed ; and to use an extreme example, supposing a large diameter cutter gets into a machine where the speed of feed is only designed for a smaller diameter, that large cutter is handicapped, because having more teeth it should take more feed, but cannot because the machine will not admit, and therefore a more effective cutter becomes defective.

It is a small point, but nevertheless an important one, to bestow care upon the roots of the teeth, keyways, and all corners, in order that they have a good radius, for it is a well-known fact that square corners will go in hardening. Mild steel sections, with ordinary wear and tear, are not permitted to have square corners.

Granted a true spindle, then it is unnecessary to have a bearing for the full length of the shank of the cutter, but a portion of its length should be backed or hollowed out, the cones top and bottom only being a good fit. For large horizontal cutters, placed two or more upon one spindle, the cones are dispensed with, parallel bearings being used, and chambering is always resorted to, as the ends only are required to be a good fit, and it is a difficult matter to lap the middle out to a standard without enlarging the ends. This recess should be as little as possible, $\frac{1}{32}$ inch in diameter being quite sufficient.

It is needless to make comparison between milling and other machine tools, or to discuss the detrimental effects of cutters being untrue, blunt, or many of the other defects which may present themselves to the practical mind. It is also difficult to estimate and compare the cost of work done by milling with that of other machines, because sometimes the milling machine embodies the work of various others. Radii, both inside and out, are formed by the cutter, &c., but where it replaces slotting or planing only, its superiority is at once recognised.

The machine itself must be of sufficient power in every respect to overcome a maximum exertion, and also perfectly rigid. In fact, too much cannot be said for rigidity. Let any part of the machine be defective, and it will rebound upon the work done. Some examples are, insufficient support of the spindles, or the table upon which the work is cramped only bearing upon one-half of one of the slides. This table should be so constructed that the work to be done never overhangs the bearing slides of the table whilst under operation of the cutter, as the author has observed it to do in some machines. Any improvement in driving gear to produce steadiness is at once adopted, and a worm-driven cutter would be an approach to perfection. The speed and feed should also admit of easy regulation for the cutters of various diameters. The feed should be adjusted in such a manner that it may be easily altered if necessary without stopping, otherwise the surface will be injured. This is done mostly by the aid of cone pulleys, having three steps, which represent, say, $\frac{3}{4}$ inch, 1 inch, and $1\frac{1}{4}$ inch feed per minute, which does not perhaps achieve the above regulation. There are also various other designs which perhaps approach more perfectly the change of feed during operation, effected generally by a frictional agency, some having an index showing the number of revolutions of the cutter per inch of feed. Transverse cuts are also maintained by feed gear, but irregular profiles are produced by the use of a cone, kept by the aid of weights against a former. The cone is necessary, because any alteration in the depth of cut can then be obtained by an adjustment of the cone either up or down. Finally, a good lubricant, copiously applied to the cutter, is a necessity.

The work done by a cutter is affected by the number of teeth, speed, and feed. The feed being directly proportional to the number of teeth, then, if the most economical speed is determined, the quantity of material removed depends upon the number of teeth in the cutter, the latter being designed with all due regard to easy production and re-grinding, the former taking into consideration primarily the stability of the machine, also the material being operated upon, its skin, finish, cubical contents removed, class of cutter, and any other factor that may present itself to the operator. The following table gives the speed and feed for various operations upon different materials.

The speed for cast iron appears to be open to the greatest amount of contention, and the author gives 45 feet per minute as that from personal experience works most satisfactorily, although from 10 to 80 feet per minute have been recorded. The difficulty lies in the fact that a lubricant cannot be used in the ordinary way, because of choking the pumps; consequently, with high speed, the temper of the cutter is soon destroyed, which is not economical.

Although the feed may be sometimes even more than that given above, the rate seldom exceeds $\frac{1}{50}$ inch per revolution, and frequently only one-half of that; and where gauge work is done, more than the proverbial roughing and finishing cuts are necessary.

The emery discs for grinding milling cutters are about 4 inches diameter and $\frac{1}{4}$ inch wide, and run from 4000 to 6000 feet, circumferential speed, per minute. It follows from this speed that any inaccuracy due to eccentricity, uneven balance, or an untrue spindle, is immensely intensified; consequently, these points must be attended to. Also the belt should be even and true; but above all, keep the wheels from glazing and free from moisture. Let them cut freely, otherwise they will soon blue-burr; finally, let the finishing cut

Material.	Object.	Description of cut.	Diameter of cutter.	Depth of cut.	Width of cut.	Speed. Feet per minute.	Feed. Inches per minute.	Remarks.
			in.	in.	in.			
Mild steel or iron.	Connecting-rods	Roughing / Horizontal	9	$\frac{3}{8}$	$10\frac{3}{8}$	45 to 50	$\frac{5}{8}$ to $\frac{3}{4}$	
"	"	Profiling ..	3	$\frac{3}{8}$	8	50	$\frac{5}{8}$ to $\frac{3}{4}$	
"	"	Finishing	..	$\frac{1}{32}$	..	60	1	
Steel casting	Hornblocks	Roughing	3	$\frac{1}{4}$ and $\frac{3}{8}$	6	36	$\frac{1}{2}$	
		Finishing	3	$\frac{1}{32}$	6	36	$\frac{3}{4}$	
"	Cross stay ..	Rotary ..	..	..	..	..	..	Ending rotary facing machine; 20 tools in a disc 2 ft. diam.
"	Motion plate		..	..	..	25	1	
Iron casting	Steam chest covers	Facing ..	6	$\frac{1}{4}$	..	45	1	Machining all round at one fixing.
"	Soft metal ..	..	..	..	..	60	..	
"	Foot or drag plate	Rotary ..	..	..	..	25	$1\frac{1}{2}$	Above ending machine.
Brass	Various ..	..	..	..	..	120	..	

holes for the crank pins, which have been bored out, so that by fixing standard mandrils in the T slots of the table, these

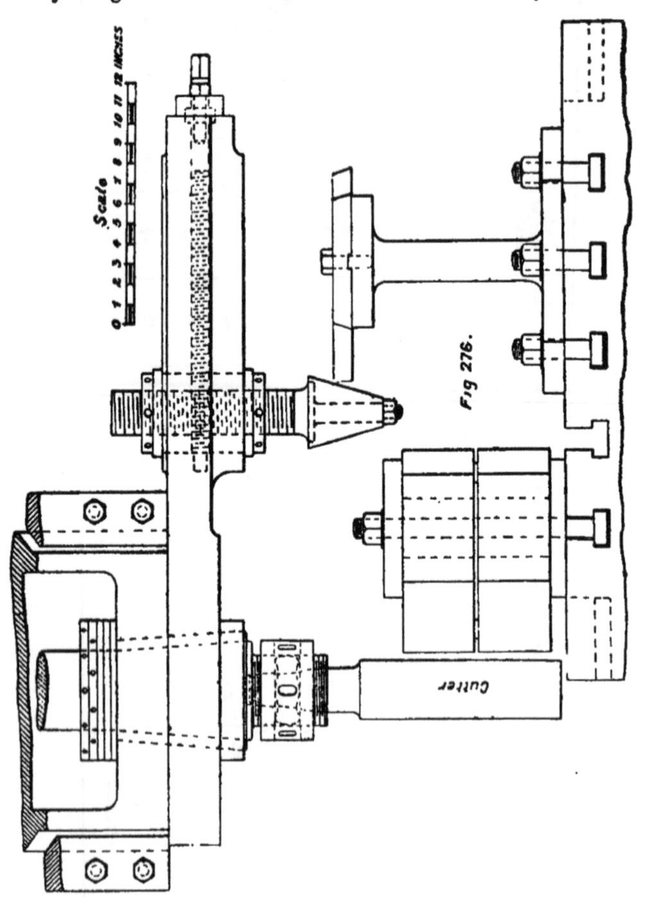

Fig 276.

rods are very expeditiously set; moreover, there is no encumbrance whatever to impede the progress of the feed when once started. Fig. 277 is a side elevation and plan of the former and its stand. After these operations have been accomplished, the bolt and oil syphon holes are drilled and rimered out, which finishes the machine work. Nothing now remains but to finish on the fitting bench, by coupling up the brasses

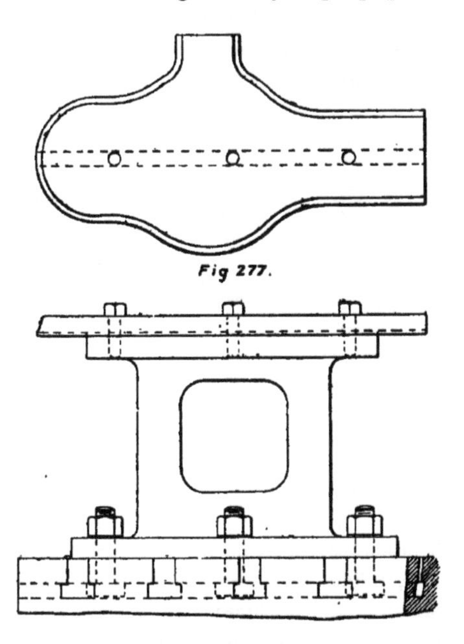

Fig 277.

and straps to the connecting rods and forcing in the bushes of the coupling rods, then all is ready for the erecting shop.

The valve rods and anchor links, Figs. 233–35, p. 140, are first milled upon each side of the boss, and then the pin holes drilled two at once by the aid of a jacket, with case-hardened bushes and a twist drill. Afterwards they are put on a 2-inch mandril in the centre of the table of a vertical milling machine and worked round the bosses.

Then they are placed upon the table of a small profiling machine, and milled upon the flats by the aid of a jacket, which is secured to the table and made specially for this work; it is of sufficient depth and width to receive the ends of two links—Fig. 278 clearly showing this. They are then dropped upon a pin fixed in the same table at each end, and secured with nut and washer and milled round the edges, cutters being used which leave all fillets and radii alike.

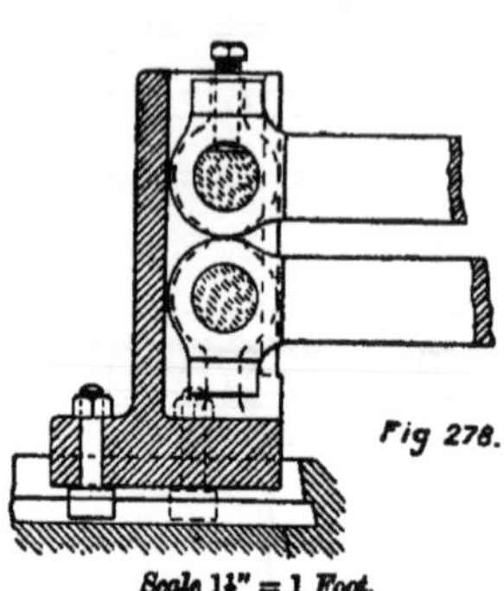

Fig 278.

Scale $1\frac{1}{2}'' = 1$ Foot.

The stirrup link, Figs. 237–239, p. 142, for the Joy motion, after being marked out is placed upon a vertical milling machine, and the outside faces milled up by parallel traverse of the table, then the pin holes are drilled and again dropped upon the milling table, and the profile edges gone over. This, as well as the swing links and all other work with sides not parallel but straight tapers, is set so that one side is made parallel with the cutter or the traverse of the table, milled, and then the other side done. After each side has been machined, it is put on to the table of a vertical cutter having circumferential movement, which mills the circular ends or round corners. The only rough portion remaining is between the forks, and this is machined by using the jacket shown in plan and elevation, Fig. 279, and section through A B, Fig. 280, which is held to the table by aid of the T slots, and the cutter passes to its work after a very rapid fixing. Finally, Fig. 281 (refer also p. 159, Fig. 253) shows, perhaps as well as any previous illustration, the ease with which a difficult job for a slotting or shaping machine is done.

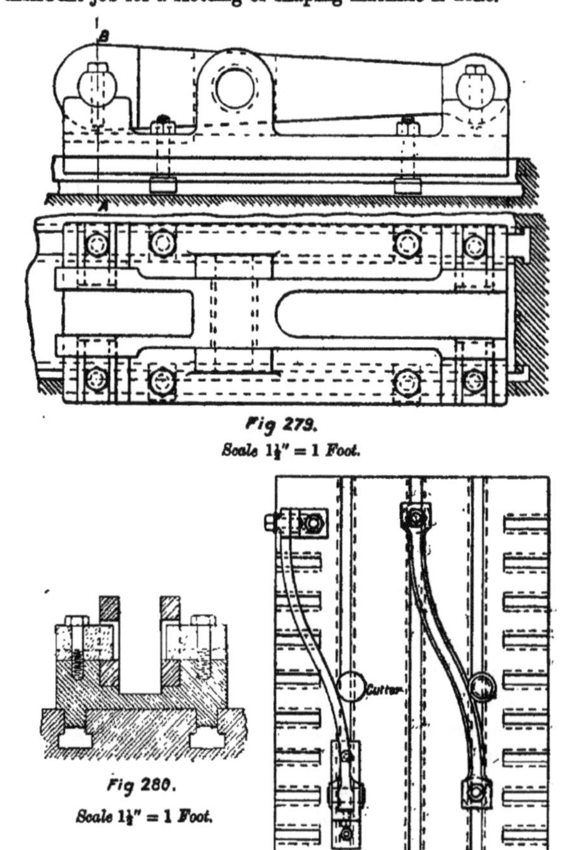

Fig 279.

Scale 1½" = 1 *Foot.*

Fig 280.

Scale 1½" = 1 *Foot.*

Fig 281.

Scale ¾" = 1 *Foot.*

Now that milling as a class of work has been dispensed with, it may be advantageous to follow the rest of the machine shop practice in a more or less precise order, such as frames and appendages, wheels and axles, cylinders and motion, &c. Fig. 282 shows the frame plate as it is received from the mill floor, also as it leaves the slotting machine, the centre lines of buffer, cylinders, axles, and the drilled clearance holes for the slotting tool. All frames are first put on to the levelling table, tried over, marked and straightened by the aid of two hydraulic jacks, which are attached separately on carriages, or conjointly on one carriage, having longitudinal movement, the jack itself being capable of traversing transversely. Dealing with frames, it has been found preferable to dispense with the thin template, as it is so liable to get out of truth by constant use, such as buckling, no matter how well it is braced with angle irons, which are always more or less in the way, and the means adopted in its place is to mark out one frame, which is a two hours' job, and

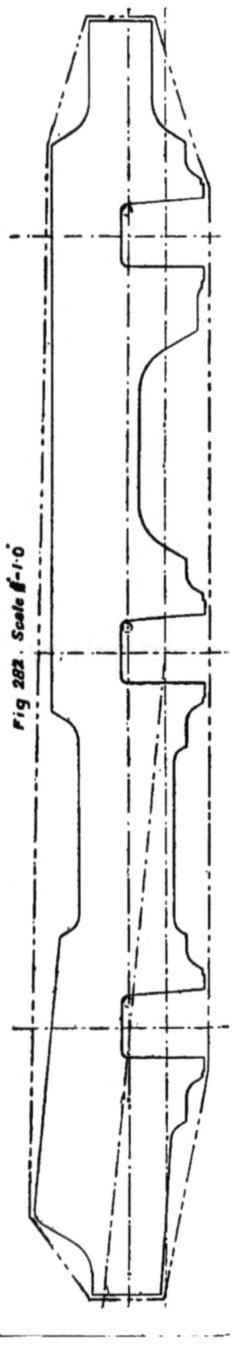

Fig 282. Scale 1/4-1·0

proceed in a somewhat similar manner to that described in the boiler shop practice, p. 7. This serves as a jacket for drilling and marking out, until the last batch is finished. They are slotted at a three-head slotting machine, four pairs—or eight frames—at once, each head having a drilling arrangement for a slotting tool clearance, the plates being first roughed out by the three parting tools, which are $\frac{7}{8}$ inch or 1 inch wide at the point, leaving $\frac{1}{8}$ inch for finishing. All bolts and cramps are slackened previously to finishing, to allow the elimination of any spring or buckle that may be held by internal strain, and released by the removal of the roughing-out pieces. They are then finished, twelve strokes per minute, and $\frac{1}{16}$-inch feed per stroke, after which they are removed to the drilling machine, which really consists of two radial arms, 18 feet 4 inches apart, the table for each being a carriage or trolley on an 18-inch gauge capable of free traverse, which enables the driller to give the frames longitudinal movement as the work proceeds, the whole process being exactly similar to that adopted for the barrel plates in the boiler shop, and previously referred to. Some frames require a set inwards at the smoke-box end, for radial-wheel clearance, and this is done on the straightening table by the aforementioned hydraulic jacks. The cross stay, motion plate, and the foot or drag plate are first machined on an ending or double-headed rotary facing machine, which will face up anything that goes between the frames of an engine. A little margin is allowed for adjustment of each head, which are 2 feet in diameter, there being twenty tools fixed in each disc, by wedge bolts. The machine is speeded to 25 feet per minute, and 1-inch feed for steel castings, increased to $1\frac{1}{4}$ inch for iron castings. All the necessary holes for bolting or riveting to the frames, slide-bar brackets, anchor links, &c., are drilled and bored by the aid of jackets, to standard, and the steel castings are sent to the grinder to be scaled. The foot or drag plate has the bearing for the tender-buffer rubbing blocks machined by an end milling or facing cutter. These blocks are case-hardened mild steel plates, 1½ inch thick. It is then planed for the various brackets, for holding such as the brake-shaft carriers, &c., everything being finished to template size. Illustrations of these castings are given on p. 60, Figs. 66–68, p. 88, Figs. 87 and 103.

The hornblocks, see Fig. 98, p. 90, are first planed for the frame seating to template, a quantity at once, at 21 feet per minute, and then sent to a vertical milling machine for machining the axle-box seating, and the space for the adjusting wedge. This is accomplished by placing the casting planed face downwards upon a jacket, which at once fixes it, and then passing a cylindrical cutter over each side, which covers the whole surface. Afterwards this is removed and a facing cutter placed upon the same spindle to cut out the groove for the adjusting wedge. It is then slotted at the bottom for the keep and also for the wedge bolt hole. The holes are then drilled to 1-inch standard, through the cast-iron jacket, Fig. 283, which has hardened bushes. It then goes to the bench, and the wedge, keep, &c. are all put together and sent in sets to the erecting shop.

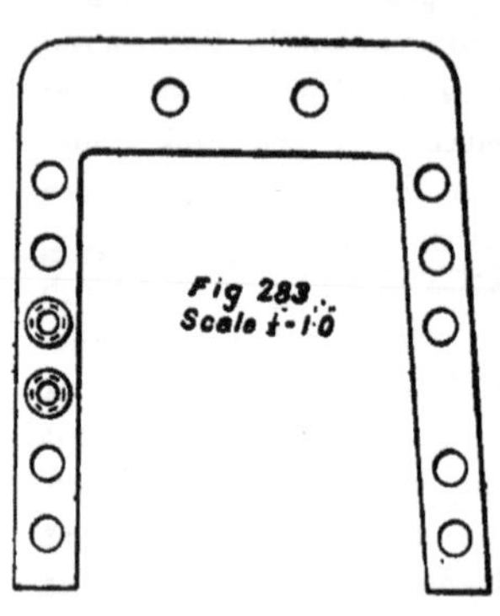

The axle-boxes, shown in Figs. 157–161, p. 112, are made of phosphor bronze. Steel castings are also used, case-hardened wrought iron being a thing of the past. The bronze boxes are first put on to the planing machine twenty-two at once, in two rows of eleven each, upon parallel strips, and secured to the tables by bolting to the T slots, for machining to gauge between the jaws for the keeps, the cutting operation being performed from the cross slide, and the feed vertical. After the keeps have been planed, they are dropped in a good fit, and the 1⅛-inch hole opened out for the mild steel case-hardened pin which carries the spring link. The boxes are again fixed on the planing machine, attached to a double angle plate, which is secured to the table, and brings them within range of the tool boxes upon each upright, besides those on the cross side, thus machining one horn seating and one face of twelve boxes, six on each side, at one operation, this method being repeated for the other seating and side. Afterwards they are taken to the bench, and the lids or caps to the oil and tallow boxes fitted on. In passing to the erecting shop, they stop at a handy little shaping machine which cuts the oil grooves in the horn seating (a much neater and quicker job than chipping); and here it may be mentioned that many oil grooves in various seatings, &c., such as slide blocks, are milled in at a quick rate by a small circular cutter. The boxes are bored six at once in a machine located near the erectors, and dealt with in that section. The speed of the table for planing is 25 feet per minute, and the feed $\frac{1}{16}$ inch for each stroke. This speed, compared with that for turning brass, appears very low, but it must be remembered that wear and tear, which takes place in a high degree upon every reversal of the table, would be accelerated by any increment of speed.

The crank axles, see Figs. 219–222, p. 130, are first put in the lathe, square-centred, cut to length over all, and rough turned on the wheel seat and journal, leaving ⅛ inch for finishing, and also a square ridge in the fillets A, Fig. 284, to assist in the adjustment of the quadrant centre plates.

They are then placed on the setting-out table to ascertain the error, if any, in the twist, which is never more than ¼ inch. Every crank must be tested for this, and the quadrants accurately set accordingly, for the finished crank depends entirely upon the initial adjustment and firm fixing of these quadrant centres. The error itself is discovered by setting level the *horizontal* sweep, then fixing a square up against the ridges in the fillets A, and marking at the top of the vertical sweep upon each side, which shows the error to be the formation of more or less than the right angle. This is then divided between the two throws, by canting the horizontal sweep in the required direction, which alters slightly the position of the

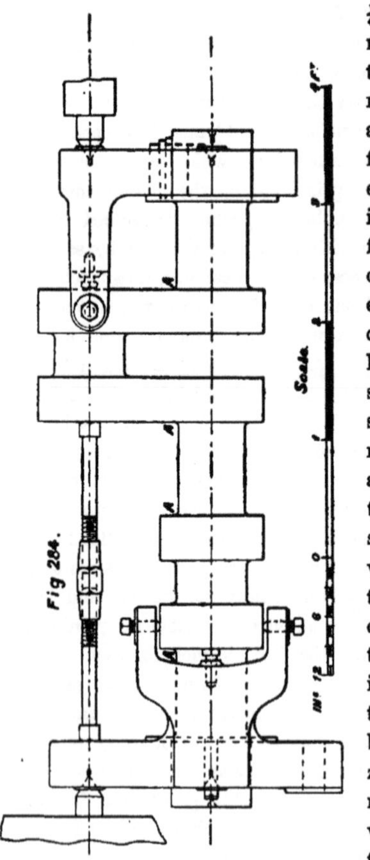

centre lines, and is finally eliminated by the slotting machine. The error is very seldom sufficient to necessitate the return of the crank to the forge for readjustment; for the twisting template is better looked after than to admit of this. The quadrant centres are then adjusted by the set screws and rigidly keyed up, care being taken that they do not move during this operation. Fig. 284 gives the longitudinal elevation and Fig. 285 the end elevation of these centres, with jack bolt complete. The whole arrangement is then placed in the crank sweep milling machine, which, *inter alia*, consists of a disc with about eighty ordinary tools, wedged in the periphery, 4 feet 2 inches in diameter,

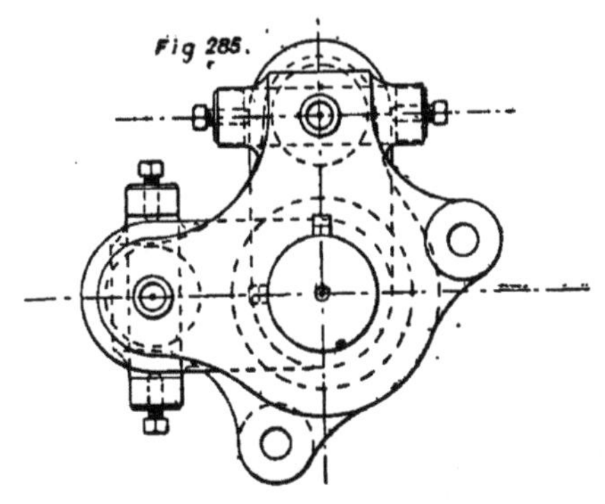
Fig 285.

and having a circular speed of 10 feet per minute. The feed can be adjusted so that as a less number of cutters are engaged simultaneously, it can be increased, and may average 6 inches per hour circumferential speed of feed, the centre of the crank pin for the radius; or in other words, the whole space may be removed in from eight to ten hours, leaving $\frac{1}{8}$ inch for finishing the crank pin. This machine has been described upon more than one occasion, and is well known. The crank is then removed to the finishing lathe and completed, four tools being used, that is, two back and two front slide rests. The webs are then slotted round, there being two tools in separate boxes on the same head, each the required distance apart, which arrangement necessitates only a 5-inch stroke, twelve per minute. It is fixed for slotting by simply dropping into a box, Fig. 286, which is fixed and secured to the tables, and has a graduated scale which facilitates any required adjustment. The straight axle, Fig. 226, p. 133, is turned

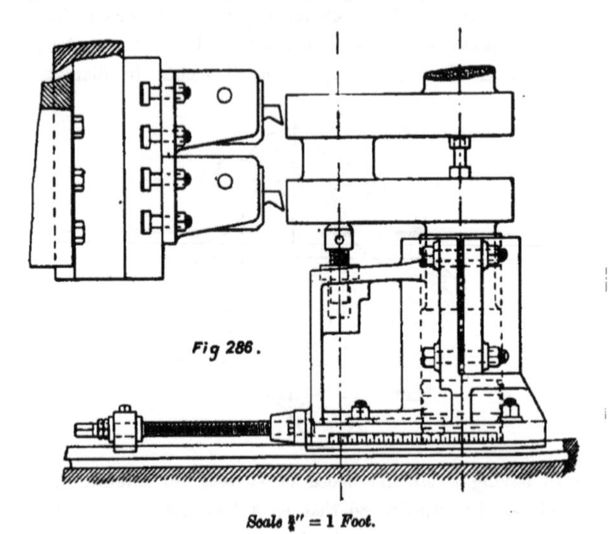

Fig 286.

Scale $\frac{3}{4}$″ = 1 Foot.

accurately to gauges, and the same speed is employed as for the pins and journals of the cranks, viz. 22 feet per minute, with a feed of $\frac{3}{8}$ inch to $\frac{1}{2}$ inch per minute. The key seatings are milled out at one cut on a specially arranged horizontal machine. They then have the wheels pressed on at the wheel press, a total of one ton to two tons per square inch on a 9-inch ram being employed. This range of pressure is not in any way due to inaccuracy of turning or boring, because every axle and wheel seat is turned and bored upon every occasion to gauge and template, as near as any operator's touch will permit; but mostly to a difference in materials, cast iron of course requiring the least, and in steel castings some wheel seats may be harder than others, and a little extra resistance in the wheel seat makes a large difference on the pressure of the ram. Afterwards the wheels are keyed up, the crank pins for the outside rods fixed, and the whole sent to the erectors.

The wheels, Figs. 91, 94, p. 88, after being annealed and dressed up, come into the shop with the gits on. They are "chucked" by the aid of ordinary dogs to the face plate, and this git is removed by a parting tool, which is made from bar steel 3 inches deep, $\frac{5}{8}$ inch wide at the top, and $\frac{1}{2}$ inch at the bottom, jumped up to $\frac{7}{8}$ inch wide at the cutting edge. While thus fixed they are rough turned all over, including the wheel seat, "topping" on the tread, and also the sides of the balance weight. They are then re-annealed, and the wheel seat is bored to the standard size and pressed on to the axle. The pair are then turned to gauge for the tires, the speed varying from 16 feet to 18 feet per minute. Fig. 287 shows an extremely handy chuck for wheels of small diameter, say bogie, &c., which is attached to the ordinary face plate. The drivers are represented at A, the bearings for the spokes at B, and adjustment under the rim at C. This arrangement admits of the following operations being performed during the one setting: boring the boss or wheel seat, turning the tread, and facing upon each side of the rim.

The crank pin seatings are then bored out at a quarter-centre boring machine, parallel and both at once. This machine consists essentially of two headstocks, fixed upon one bed at a suitable distance apart, and in each is the gearing for driving the boring bar. The wheels and axle are placed between the centres, one crank pin seating below in a vertical centre line, the other seating will then be either leading or trailing. Between the wheels are two other standards or frames, for supporting the other end of the boring bar in a pedestal; this for the vertical seating is

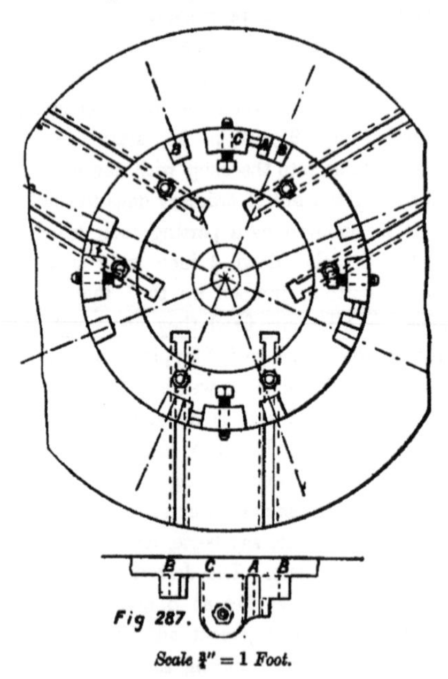

Fig 287.

Scale ¾" = 1 Foot.

simply arranged for upward or downward traverse only, whereas the opposite can be transferred as required, leading or trailing, with a certain range of horizontal traverse, the boring bar being also capable of similar rearrangement. The wheels are now ready for the tires, which have been bored and grooved to template at 10 feet to 20 feet per minute. These are put on by expanding, Bunsen jets being the heating medium employed, which is simply a perforated tube encircling the tire, and giving to the whole a very uniform heat. The ⅞-inch tapping holes are then drilled through the rim for the securing bolts by a double-headed drilling machine, a smaller hole being drilled into the tire 1½ inch deep for the end of the set screw, it being turned down accordingly.

The cylinders—Figs. 53–60, p. 50—are first placed on the setting-out table, centre lines drawn, and the whole casting checked over. They are then placed on the planing machine, bottom upwards, set to the centre lines, and the edge upon each side is planed. The tool is then set to a 2-inch gauge from this edge, and the drain cock nipples planed up, which afterwards act as the foundation for the setting for future machine work. The front end is then planed to gauge, from the centre of the steam ports to the face, also at the same setting the angles for the smoke-box tube plate and the door plate are cut, the cylinders being fixed at the required angle, or canted, if they are so placed between the frames. They are then planed upon each side at one operation, for the frame seating, to standard width, the same setting being also utilised for machining the port face and the smoke-box steam chest cover face. The planing operation on each is one good sliding cut, with a tool having a large curved point, broad shavings being removed with a good surface finish, at from 18 to 20 feet per minute. The next operation is to fix them upon a table, which is on the floor level, within a jacket of standard width, which is really the frame of a machine carrying four boring bars, pitched out exactly to the centre of the piston and valve rods, which enables the stuffing-boxes to be bored out at one operation, without any setting whatever. The boring mill is arranged very similarly, each cylinder being bored simultaneously at 11 feet per minute, with $\frac{1}{16}$ inch feed for roughing, and ⅛ inch finishing per revolution. They are then ready for the steam and exhaust ports, and the bearings for the guide bars to be either slotted or milled to width or thickness. At the drilling machine the drain cock holes are the first drilled, and then the bolt holes for securing to the frames. These holes are 1⅛ inch standard, and are drilled through a jacket, which is simply a cast-iron plate with standard bushed holes, the whole arrangement being set at once by lugs, which are cast on to the plate, coming flush up against previously machined portions of the cylinder, whereas the jacket for the steam chest cover

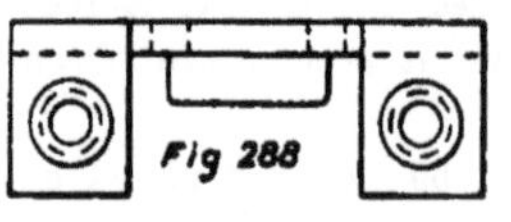

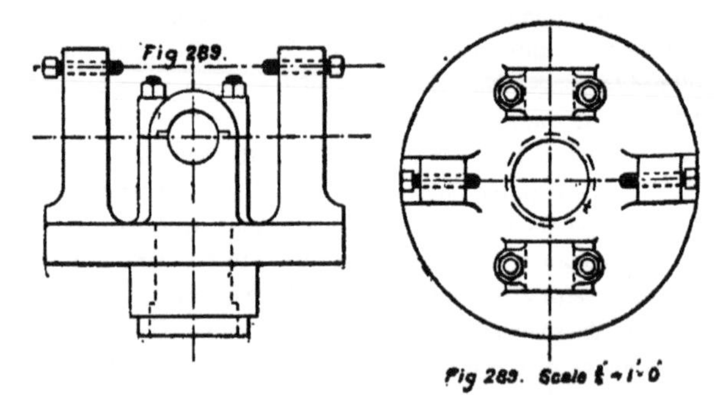

studs has centre lines, which are set to those on the cylinder and then cramped. Fig, 288 shows the drilling jacket for the holes in the guide bar brackets at the cylinder end. The stud holes are drilled, tapped and studded by a lightning tapper, at 160 revolutions per minute, with 1½ inch downward traverse per minute, the steam pipe holes and seatings being bored out at the same setting. They then pass on to the cylinder fitters, all covers are adjusted, glands put in, valves and spindles tried, and all nuts put on, those outside being case-hardened, and those in the smoke-box being brass box nuts.

Each piston head is chucked and the cone bored. It is then fixed on a cone mandril, screwed tight up and placed between the lathe centres for turning the periphery and ring grooves. The rods are turned, also the cones at each end, to standard gauges, and then sent to the cotter-hole machine,

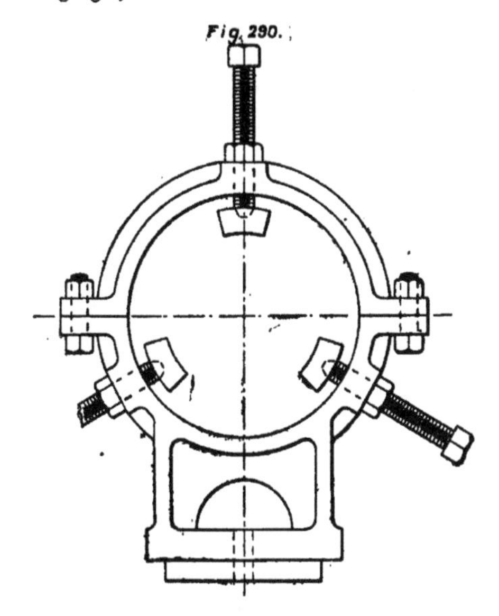
Fig. 290.

which is a double headed slot drilling machine. The cross-heads, Figs. 89, 90, p. 89, are faced up on both sides, and bored out for the gudgeon by cramping to the face plate. They are then placed in the special chuck, Fig. 289, and rough turned, afterwards the stay, Fig. 290, is placed in position, and the cone bored and finished. The centre of the loose headstock is brought up to the cone, the stay removed, and the outside is finished. This stay is a useful appendage, and can be brought in for numerous jobs, both standard and emergencies. It only remains now for the edges of the cross-head to be milled.

The slide bars are planed all over, four at once, set in a special bracket attached to the table of the machine, at a speed of 18 feet to 20 feet per minute. They then go to a small slotting machine for the clearance to be slotted and to

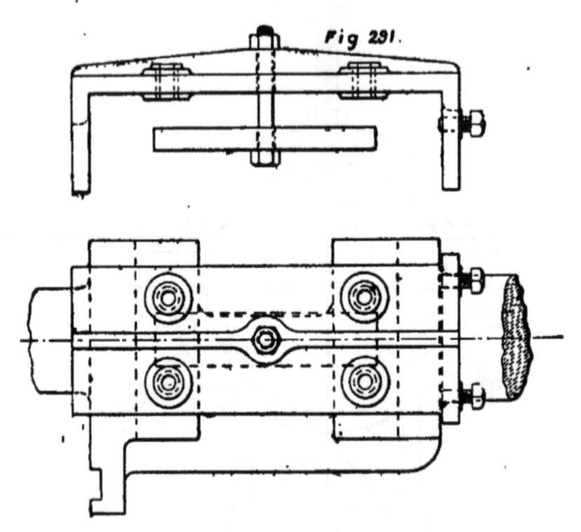

be cut to length, afterwards the holes are drilled, and then finished by glazing. The slide blocks are also planed, a quantity at once, the recess for the white metal, and also the oil grooves, being milled at a rapid rate.

The rods and the rest of the motion in the ordinary course of events would have followed on here, but these were dealt with under milling and its operations, so that we must now pass on to the reversing shaft and radius links, and conclude with this class of work. The reversing shaft,

Figs. 95, 97, p. 90, is first set out, then turned up in the middle and at the ends, after which it is slotted out to receive the radius links, which are put in a driving fit, being pressed by a little mechanical device, and secured by two $\frac{5}{8}$-inch set screws. The jacket for drilling these holes is shown in Fig. 291. The radius links, Fig. 236, p. 141, are first planed upon the back side, and then fifteen of them fixed upon a lathe face-plate, forming a complete circle of 3 feet $7\frac{1}{2}$ inch radius, and turned out to receive the slipper blocks, then faced up to the right thickness, and afterwards

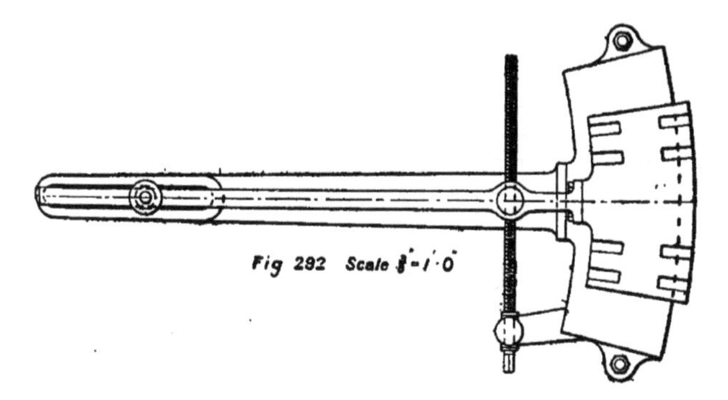

Fig 292 Scale $\frac{1}{8}$" = 1′ 0"

slotted to length and for clearance upon the flats to fit the reversing shaft. After thus finishing they are case-hardened all over, then the rubbing surfaces ground up in a vertical grinding machine, being fixed on the radius grinding table, Fig. 292, which is attached to the machine when required, and is capable of receiving a variety of articles of this description. The slipper blocks of phosphor bronze, Figs. 151 to 156, p. 110, are turned all over upon the same face plate, twenty-four forming the radius, or sometimes they are milled up, the radius being produced by a special device.

Fig. 293 is a boring jacket for opening out under a radial arm or ordinary drilling spindle, the dummy gland for the brass bushes in the front steam chest cover of engines having the steam chest between the cylinders. These jackets, jigs or rigs, are applicable to every repetition operation, whether turning, planing or drilling, and the illustrations given are a mere tithe of the host. They save time in fixing work on the machines, labour in setting out, and are certainly conducive to accuracy.

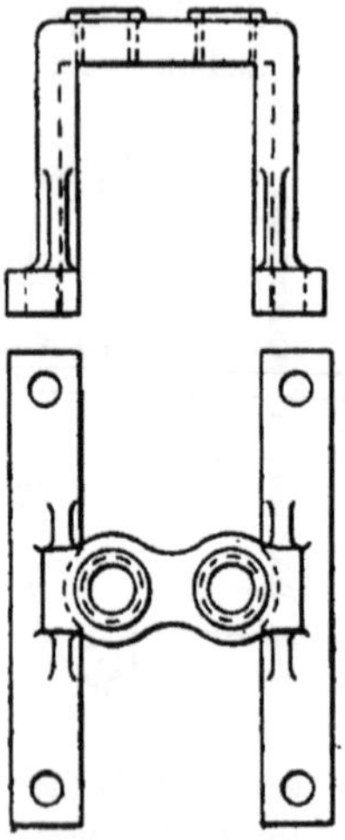

Fig 293 Scale ¾"-1' 0"

Brass finishing and its requirements are all concentrated in one portion of the shop, under one chargeman, who is responsible to the foreman. The appliances consist of lathes, which are chiefly capstan and chasing, suitable machines with fly-cutters for nuts, emery bands about 1½ inches wide, glazing wheels, "buffs" for polishing, and cock grinding machines. The buffs are generally squares of calico, a number put together, the diagonals placed to form, as it were, a many-sided figure, which when revolved at a high rate, and a little flour of emery used, produces a good polish in any corner or intricate casting after it has been first passed over a circular brush, which also revolves at a high speed. These machines are placed in the order they are mentioned, so that when all work reaches the bench there is nothing to do but fix together and test. The only proposed illustration of work done of this class is the injector, which is fully shown and described on p. 115, and by reference to those illustrations the details of the manner of work will be facilitated. It is actually an intricate piece of lathe work, requiring considerable skill and patience, as every one of the cones in the cone chamber is fixed in steam and water-tight, as well as the seating of the wing cone 2716, upon the top of the combining cone, 2714. These cones, which can be easily picked out on p. 118, are first dealt with by chucking in the lathe and boring out. They are then accurately finished by the group of boring bits, Fig. 294, which the reader can easily identify with each of

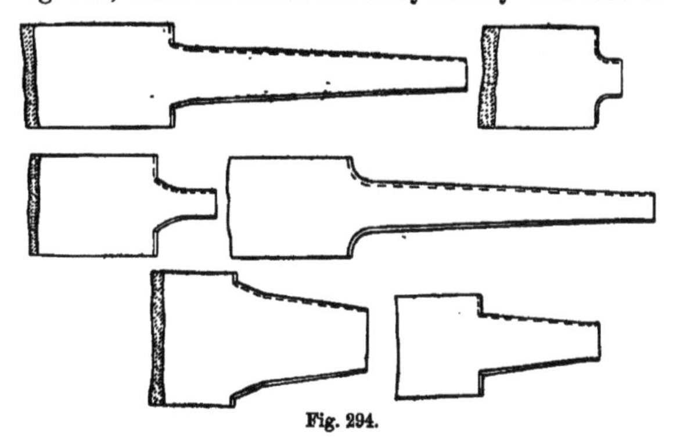

Fig. 294.

the cones in Fig. 187, p. 116. These bits are carefully turned to the standard size, then they are backed to give clearance to the cutting edges, and as the quantity of material they have to remove is very small indeed, they last an almost infinite period. The outsides of the cones are turned by chucking them from the inside on lead chucks.

Meanwhile the main casting has been set out, and the flange for forming the junction with the boiler has been planed. The seatings and stuffing-boxes for the stop valve 2719, and the steam valve 2718, are bored by chucking this flange up against the face plate. The water and overflow cock seatings 2725, are bored by chucking the same flange to an angle plate attached to the lathe face plate. The next operation is to bore out, thread and face the seatings for the various cones in the central chamber. This is done first by chipping, filing or otherwise making even and circular a bevel where the cap 2711 enters, then fixing on the chuck or centre plate, Fig. 296, to the enlargement for the overflow union, and making the top edge parallel with the planed face of the flange. This plate has three centres fixed in, corresponding to the central cone chamber, back-pressure valve seating, and the ejector steam union, shown at A, B and C.

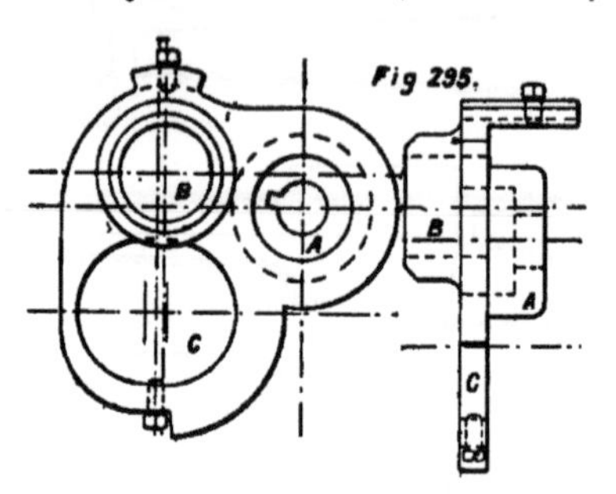

Then it is placed between the lathe centres, that at the loose headstock end being a cone to fit the prepared opening of the cap seating 2711. Then the end of the casting 2710 is simply faced and turned up, so that it will fit a cone plate and run true in it, enabling the inside of the chamber to be got at. After it has been set in the cone plate the cap seating 2711 is threaded, and then a longer tool put in, and the seating for the cone 2713 is bored and threaded, and also that for the combining cone 2714. Then a guiding plug is screwed in flush up against the seating for 2713, which extends to the face of the seating for 2724, and a boring bit is passed through, which finishes the seating for the steam cone 2715.

This chamber is now finished, and the casting is taken out of the lathe, Fig. 296 remaining on, the guiding plug is removed, and a bolt is inserted in its place, to which is attached the centre plate, Fig. 295, which is adjusted by the set screws, the projection A entering the casting flush, and B the cone plate, which enables the back-pressure valve seating to be faced. C is a clearance space for the union with the ejector steam pipe, which is attached to the injector on the left-hand side only of the engine. The back-pressure valve seating 2717 is that subject to the most wear and tear, and many old injectors would have short lives but for bushing, which has to be carefully and accurately done. In all new injectors provision is now made for an increased life of this seating, by extending the height of the casting at this point about 6 mm. or 8 mm., which, after wearing to the usual limits, can in the end again be bushed. The speed for all brass turning is 70 feet to 80 feet per minute.

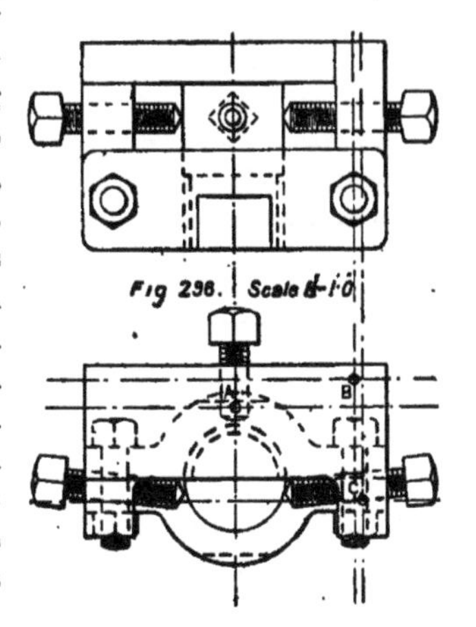

Fig 296.

All threads of every description are produced by Barrow's patent screwing machine, with one single exception, the reversing screw, which is still cut in the lathe. The following is the range of work for the largest machine:—Brake screws 1¾ inch diameter, ½ inch pitch, square thread; shackle screws, 1½ inch diameter, ⅜ inch pitch, round thread; damper rod screws, 1 inch diameter, ¼ inch pitch, square thread; and all sizes of bolts between 1¼ inch and 2¾ inches diameter. A second size screws all those from ⅞ inch to 1¼ inch, and these two machines will also, by means of plain dies, take a surface cut of the material before screwing, besides cutting and forming their own dies, by the aid of an end milling cutter, tap, and a taper milling cutter. A third size produces those below ⅞ inch and nothing less than ⅜ inch. The speed may be taken for, say 1 inch bolts, at eighty revolutions per minute. All sizes of nuts are tapped at the well-known quadruple tapping machines.

It was considered convenient to give the various speeds and feed when dealing with the individual operations, but for a compact reference the appended table may be useful, as it represents the average practice. It should be studied in

TABLE OF SPEEDS, FEED, &c.

Material.	Object.	Operation.	Speed. Feet per minute.	Feed. Inches per minute.	Remarks.
Mild steel	Crank and straight axles	Turning	22	⅜ to ½	
Steel castings and higher grade	Wheels and tires	Do.	16 to 18	..	Feed according to grade.
Cast iron	Cylinders	Boring	11	..	1/16" and ⅛" per revolution.
Brass	Finishers generally	Turning	70 to 80	..	
Mild steel	Various	Twist drilling	20 to 30	1 to 1½	
Cast iron	Cylinders	Drilling, tapping	40	1½	
Steel casting ..	Hornblocks	Planing	21	..	
Mild steel	Slide bars	Do.	21	..	
Brass	Axle-boxes	Do.	25	..	1/16" per stroke.
Cast iron	Cylinders	Do.	18 to 20	..	
Mild steel	Frames	Slotting	8	¾	
Do.	Inch bolts	Screwing	20	..	

conjunction with the conditions, and all-round difficulties which turn up in every machine shop. It is certainly possible to work at higher rates in some cases; but then the questions of tool economy and wear and tear of the machines come into consideration.

SECTION VI.

ERECTING.

We have well-authenticated instances of rapid erecting, it is not the author's intention to describe one of th From 40 to 45 hours is a reasonable time for erecting eng of the class shown in the sectional drawing. This may accomplished by a system which gives each chargeman th pairs of frames under his care, and then by a judici manipulation of his men, engines may be built very m cheaper than if there was only one set of frames for him superintend. It also must be understood that every ot department of the works is at least the greater part, if the whole of one order in advance of the erecting. T accuracy and system maintained in these other departmen is also a factor much in favour of cheap and rapid erecting which is supplemented by like care in this shop, introducing anything to minimise time or labour, and keeping all templates and gauges up to standard.

In the following figures, some of the necessary helps to erecting may be seen :—Fig. 297 shows two forms of a bottle-jack stand, upon which the frames are erected; Fig. 298, the standard distance stays, which are placed between the frames in suitable and convenient places; Fig. 299, the bit or rimer for opening out the holes in the frames for the cylinders, &c., by the aid of the Stowe flexible drilling shafts; Fig. 300, a light template cylinder cover, which by the help of a 3-inch piece of tube 8 feet long, in conjunction with Fig. 301, is used for setting the slide bars; Fig. 302 is a light template

[To face page 224.

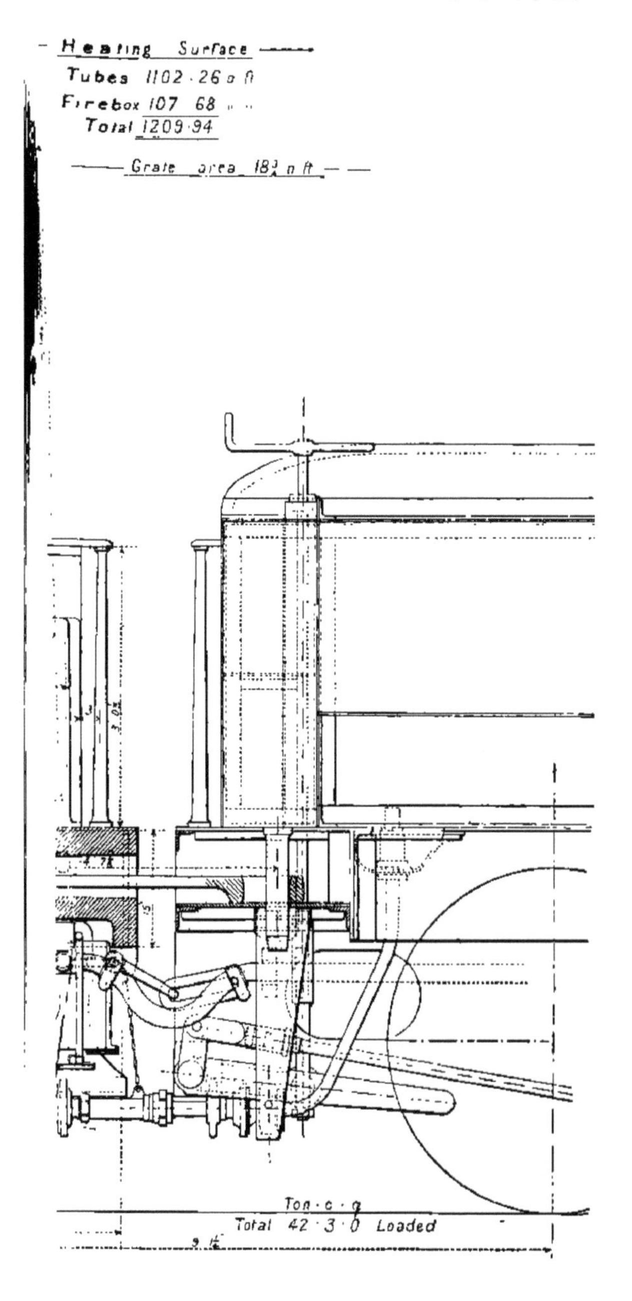

used for setting the reversing shaft brackets; Fig. 303 is a template of the standard journals, to which all the axle-boxes

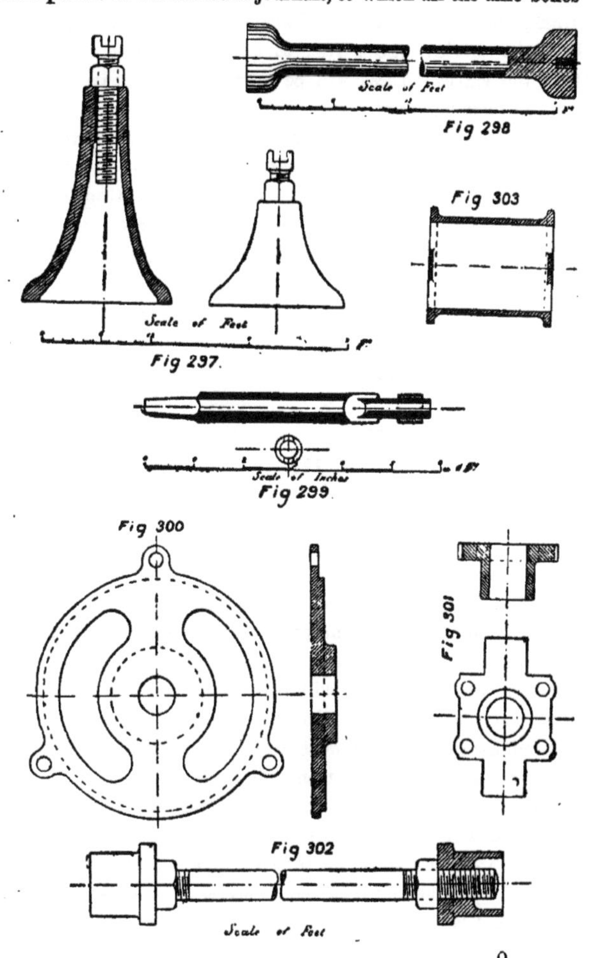

are bedded; this saves hunting up trestles and getting roller pins to slide, and eventually lifting a heavy box on to a journal; and Fig. 304 is a half-portion of the cradle upon which the driving wheels rest, to facilitate turning when setting the valves. These will be referred to again, in due course, in their proper places.

The progress of erection is illustrated by the four Figs. 305–8, each of which shows the work done at the end of its period; so by tracing the figures representing the second, third and fourth periods, and placing them all upon the first, an almost complete drawing of the engine can be formed. Those portions not shown are the last to go together, so that the whole work done in the last period can be gathered from the sectional drawing, which shows the engine complete for trial. These illustrations have the advantage of showing very concisely the work accomplished during a certain stage, which is also neither cramped, confused nor minimised by that done during any other period. Each of these figures will be described separately, and the author will take each portion of the work upon the supposition that he is the only mechanic employed, assisted by an apprentice and necessary labourers; whereas in actual practice many operations go on simultaneously, in fact as many as possible so long as the pit is not crowded with men, and they do not hinder each other.

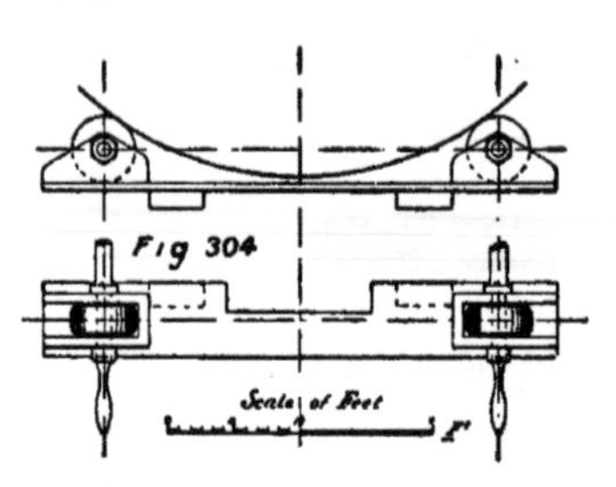

Fig. 305:—The frames are received in the erecting shop with all sharp corners filed off and all necessary holes tapped. They are first laid upon trestles, and the various centre lines

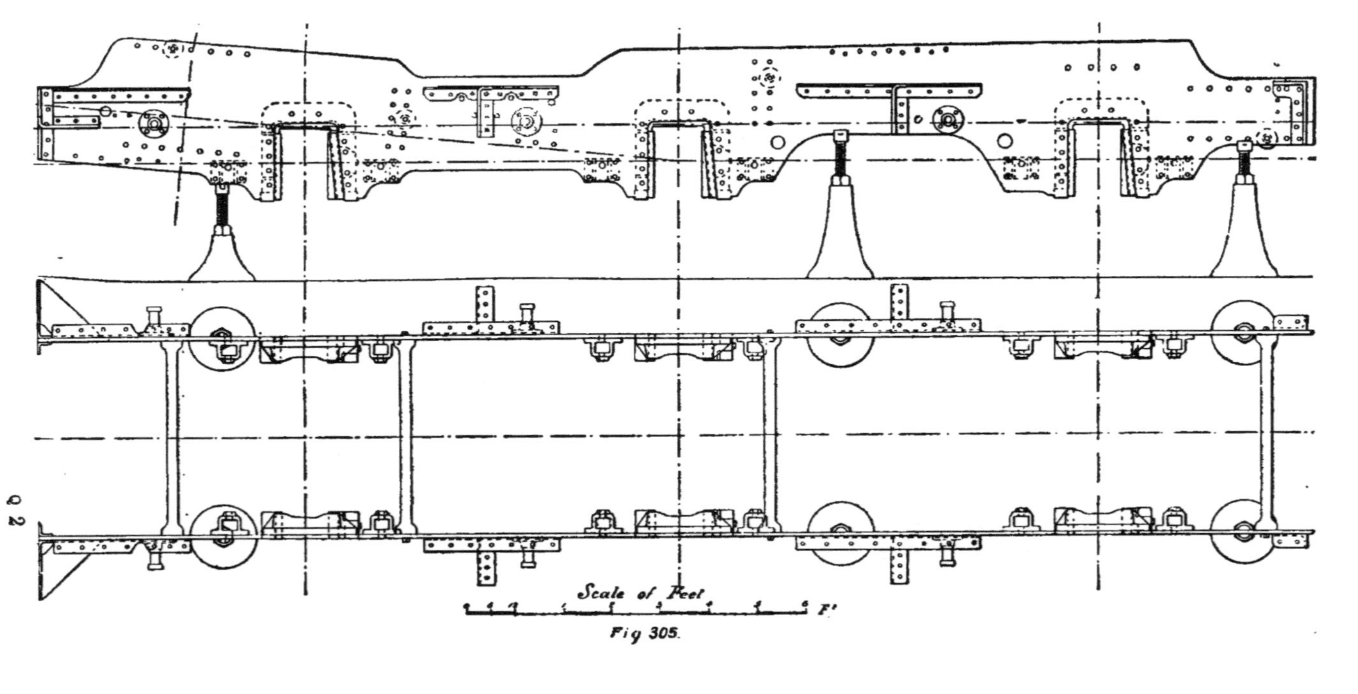

Fig 305.

set out for brackets, stays and motion. They are then reared on to the stands, Fig. 297, the first one being cramped to a trestle, or dropped into fork stands, and then temporarily stayed to the other with the standard cast-iron stays, which have been turned up at the ends to dead length, Fig. 298. By the aid of the bottle-jack stands, the frames can be adjusted to height and set square with each other by diagonal trammelling, from the centres upon the top of the frames, over the driving and trailing horns, and also set level, length and crossway, by a straight-edge and spirit level. A thin cord is also placed round them at a given distance, to prove that each frame is perfectly straight along its whole length. This cord is not shown in Fig. 305. All platform angle irons, gussets and other supports, which are clearly shown in the figure, are bolted up, the holes opened out by the flexible shaft, and afterwards riveted up by the hydraulic riveter; or, as in the case of supports, cold rivets are put in a driving fit, which are the most trustworthy, especially for the motion plate. The horn blocks are then fitted into the horns, and the holes in the frames opened out. It may be beneficial to remind the reader that these and all other mountings for the frames, including cylinders, &c., have been drilled to standard, consequently they act as drilling jackets to the frames. The horn blocks are secured by turned bolts, a driving fit. The inclination of the cylinders is 1 in 10.

Fig. 306:—The temporary angle irons A with their adjusting screws are fixed, and the cylinder casting then lowered on to them. The templates B and C are then placed at the front end and in the driving horn, at given distances from the top of the frames, by the aid in the latter case of a straight-edge across the top and a T square. A long-distance template is then passed through the cylinders to the centre of the driving horn, which gives the distance from that point to the face of the cylinder. Piano wires are

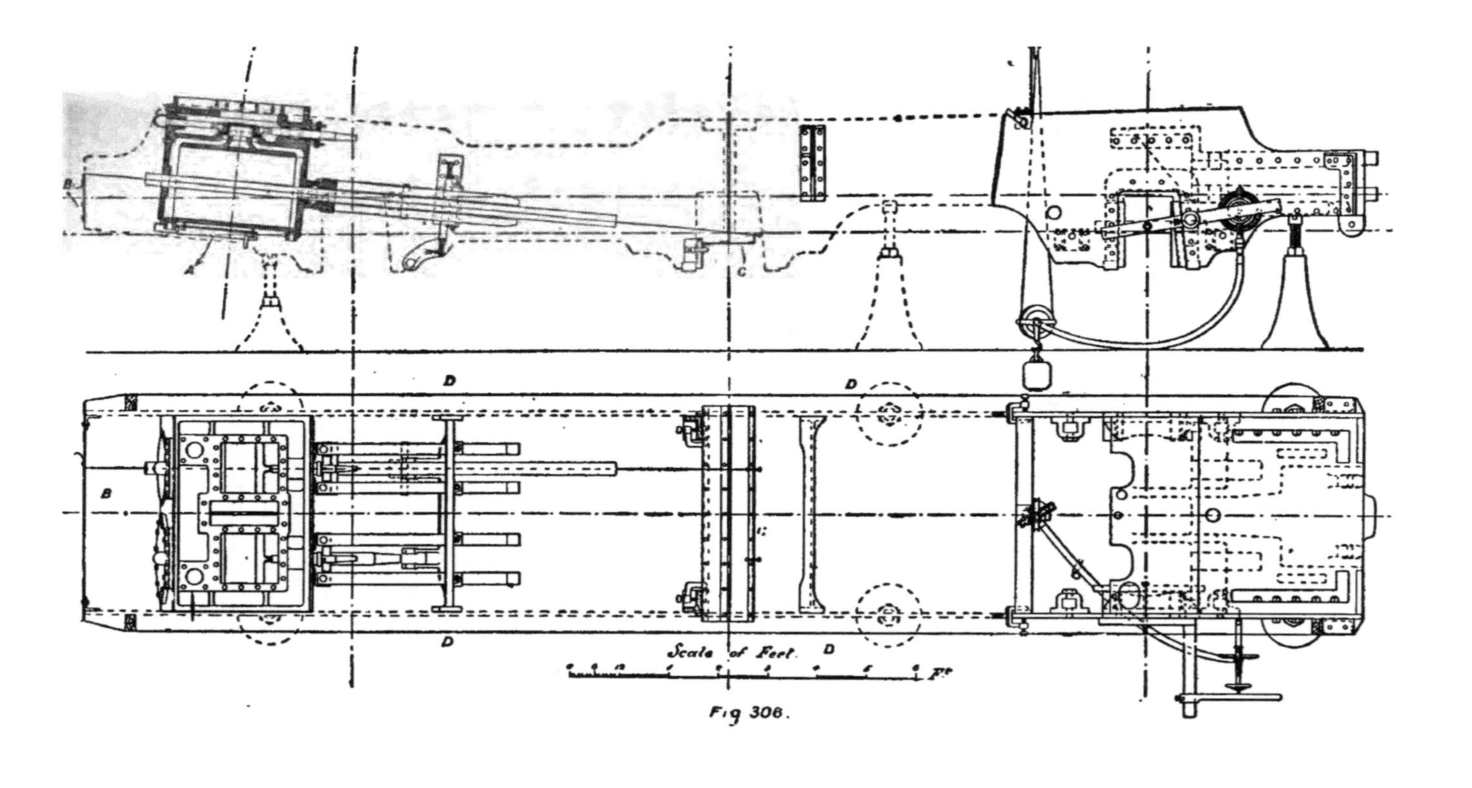

Fig. 306.

then stretched from B to C, at given distances from the inside of the frames, which also fixes the centres of the cylinders. On the template C are thumb-screws, by which the wires are made quite taut and straight. The cylinders are then set to these lines, and securely fixed by temporary bolts. The motion plate is then fixed to distance by a template, which fits in the bolt holes of the lugs on the cylinder for the slide bars, and also in those upon the motion plate. It is set by the aid of a straight-edge across from the lugs on the stuffing-box and the brackets on the plate. The template covers, Fig. 300, are now fixed to the front end of the cylinders, and the slide bars set by passing a 3-inch tube through, about 8 feet long, which has a bearing in the cover and gland. The template of the slide blocks, Fig. 301, must work up and down quite freely, but at the same time it must be absolutely free from all shake. The use of this tube and template renders the operation a certainty, and is fully appreciated by the operator after a trial. A straight-edge smeared with a very thin film of red lead is also tried over the whole set to get them square, or in the same plane with one another. There is a suitable liner placed under every bearing of the slide bars, to allow for taking up wear and tear. The cross or frame stay is then set to a centre line, which has been fixed at the right distance from the driving horn. The foot or drag-plate is set level with the hinder portion of the frame, flush and square with the end. All having been securely fixed with temporary bolts, it is then the foreman's duty to pass the whole, as he must be absolutely certain that everything is correct before anything else is proceeded with. He first walks round the frames, passing his gauge at intervals between the frame and the cord D, to ascertain the correctness of the straightness of the frames, or that they are parallel. Next, he tries if they are level, both longitudinally and transversely, also if the diagonal measurement between the driving and trailing horns is correct. He then ascertains that the templates B and C are fixed at their proper distances from the top of the frames. The face of the cylinder has been planed to a given distance from the centre of the ports, and a template has been made to ascertain the correctness of this, so that the next operation, verifying the distance from the centre of the driving axle to the face of the cylinder, shall not be nullified by any mistake on the part of the planer. He next tries the distance of the piano wires from the inside of the frames, which proves that they are parallel, also that the distance from the centre of the cylinder to the planed frame seating of the casting is correct. The insides of the cylinders are then gauged from the centre wire at the front end, also at the stuffing-box, and finally a spirit level is placed upon the smoke-box tube plate seating. All the holes are then opened out by the rimer, Fig. 299, and the Stowe flexible shaft, which is shown in position at the back end of the frames. Turned bolts or rivets, as the case may be, are then driven in and all secured. The platform edge angle irons and plates are then fixed and riveted by hydraulic riveters. The slide valves and buckles are then placed in the steam chest, a $\frac{3}{16}$-inch liner being put into the front port and the edge of the slide valve pushed up against it; this position is then indicated upon the valve spindle by trammelling a short length from the back of the cylinder casting on the spindle. This liner is then placed in the back port and the edge of the valve drawn up against it, the same trammel length marked upon the spindle, which gives the position of the valve at the commencement of the stroke or lead. The pistons and rods are then put in, and the "bump" or clearance marked on the slide bars for each end, and then everything is boxed and bolted up, a temporary cover being placed over the exhaust.

Fig. 307:—The work of this period is commenced by fixing

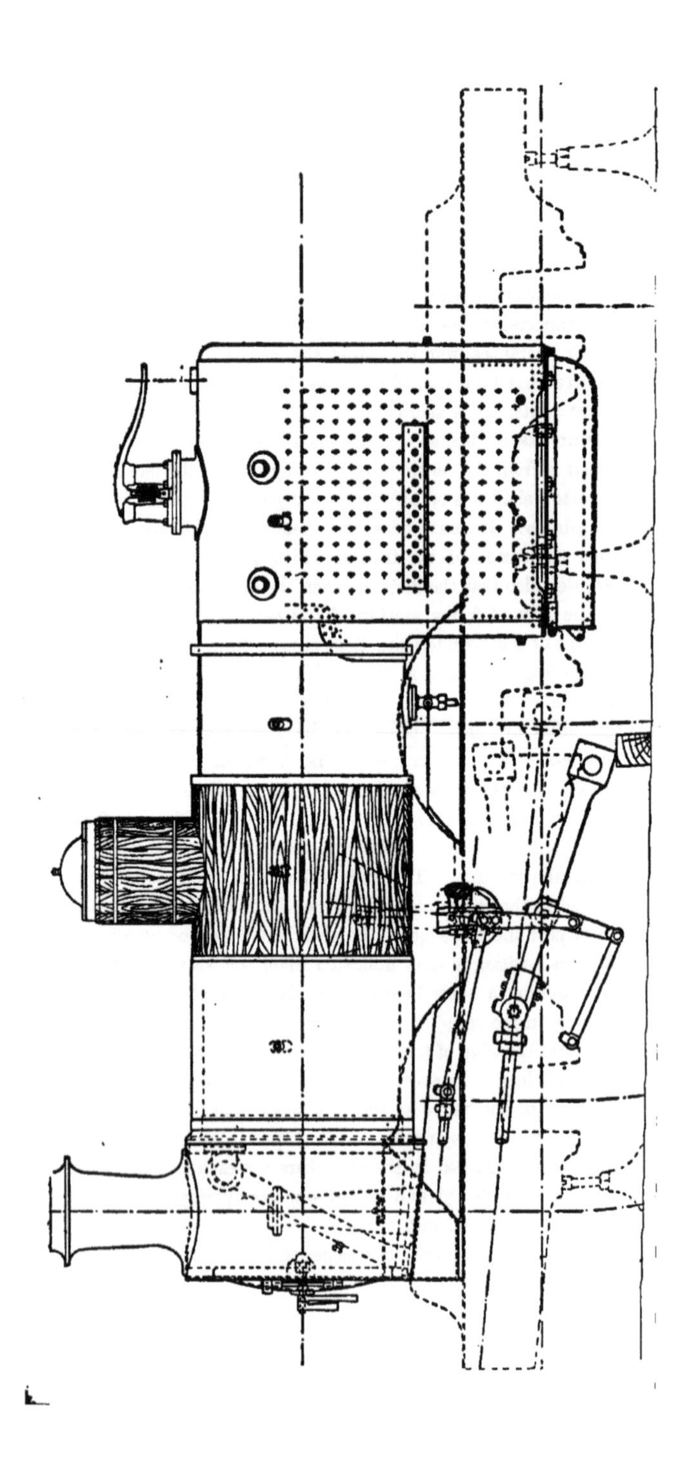

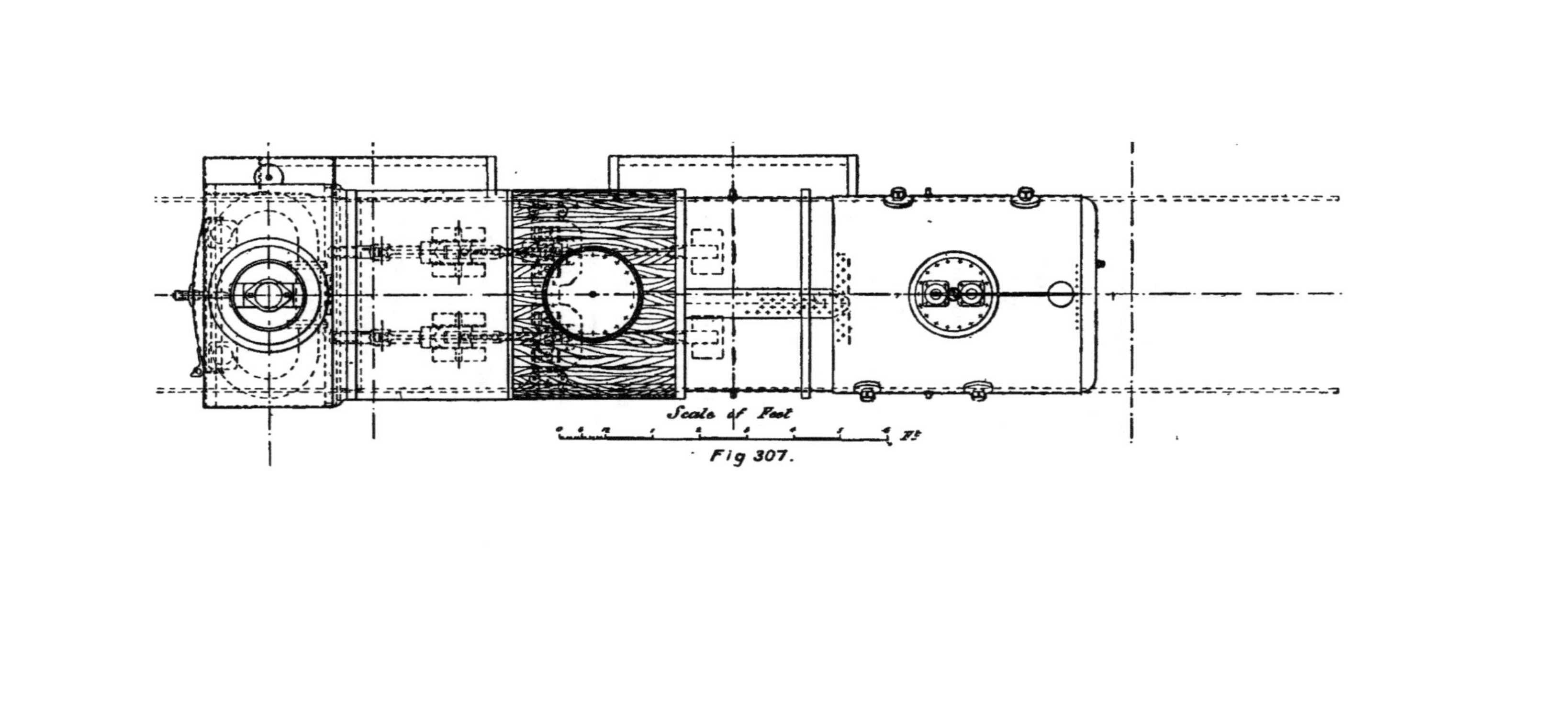

Fig 307.

the brackets for the reversing shaft, which is only temporary, the final fixing taking place after the setting of the valves, because this shaft bears a most important relation to the position of the valves. They are bolted in their places temporarily, being set by gauges from the centre of the driving horn and from the top of the frame. The template, Fig. 302, is used instead of the shaft itself, because of the great weight of the latter, and consequently if the template is a good fit, square and turns freely, the shaft can be depended upon to be the same. The connecting-rods are then coupled up to the cross-head, the big ends (minus their straps) resting on baulks of timber, or better, slung from the top of the frames. The reversing shaft is then lifted into its brackets and the rest of the motion is fixed and pinned up. The boiler which has been mounted, see p. 34, is now placed in position and secured at the smoke-box end by nine $\frac{7}{8}$-inch bolts, turned and of driving fit, the tube plate resting upon its support, which is a portion of the cylinder casting. The fire-box end is fixed by the expansion angle iron, Fig. 38, p. 24, and then the distance of the centre line of the boiler is gauged from the top of the frames by a standard gauge.

The accuracy which has been attained in constructing the boiler is such that there is never any necessity for it to be lifted out of the frames because the expansion angle iron was not square, or that the tube plate would not drop on to its seating. In works where these errors occur endless trouble is caused, because the angle iron, if not square, must be chipped upon the under side, from nothing at one end to the required amount at the other, to bring it true. The tube plate must be warmed by heaters and set back or forward, as the case may be. Frequently one or two trials and various chippings have to be made, which in the aggregate make up a loss of time at the least estimate of ten or twelve hours, one of our periods or thereabouts, and then finally it is not a good job. The boiler is then lagged with red deal dipped in limewash, which is as good as yellow pine painted with asbestos, and clothed with sheets 14 I.W.G. In the meantime the fire bars have been put in, and the ashpan cottered up to the foundation ring. The smoke-box is then built and riveted up, all rivets being counter-sunk, chipped, and filed flush with the outside case. The steam pipes are then fixed, also the blast pipe, the centre of which is a given distance from the front of the tube plate. This forms the centre of the chimney, which is set square with the boiler and plumb with the centre of the exhaust. The leading splashers, which are in one casting with the leading sand-boxes, Figs. 69–72, p. 64, are then adjusted, and also the driving splashers. The work to be done at these is very little; excepting the fixing and bolting down, little or no fitting is required. During the fixing of the wood lagging the axle-boxes are bedded, also the big end brasses, the template, Fig. 303, being used for the boxes, and then each of these is mounted on the axles. The axle-boxes are bored out a quantity at once, on a machine with horns, the boring bar of which slides out. They are dropped into these horns, consequently when bored the centre of the radius of the crown is equidistant from each horn seating. They are then tried up in the horn blocks, simply to adjust the horn wedges.

Fig. 308:—The wheels and axles which have been mounted with boxes, big ends, springs, &c., are now put under, the horn stays and big ends coupled up, and the cradle, Fig. 304, placed under the driving wheels. The engine is now ready to have the valves set. It has been previously mentioned that the valves have been boxed up, but their position is indicated upon the valve spindle at the point of lead, also that the reversing shaft has been only set temporarily, and that the "bump" or clearance has been marked on the slide bars.

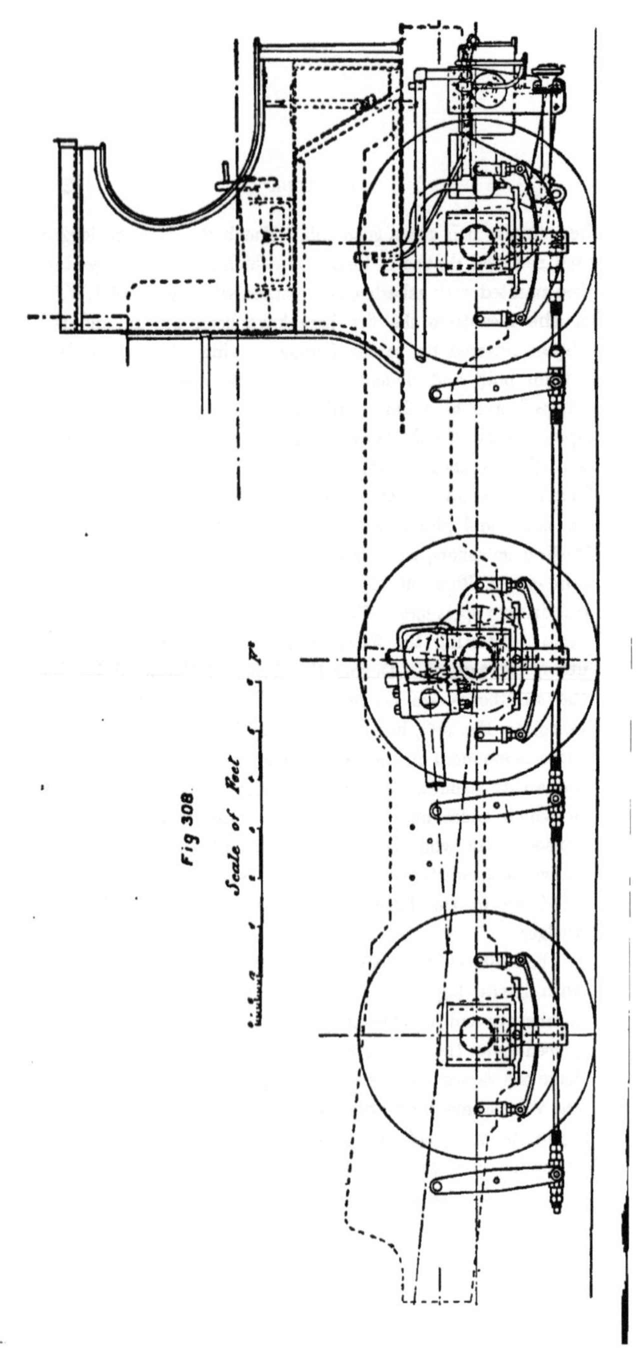

Fig 308

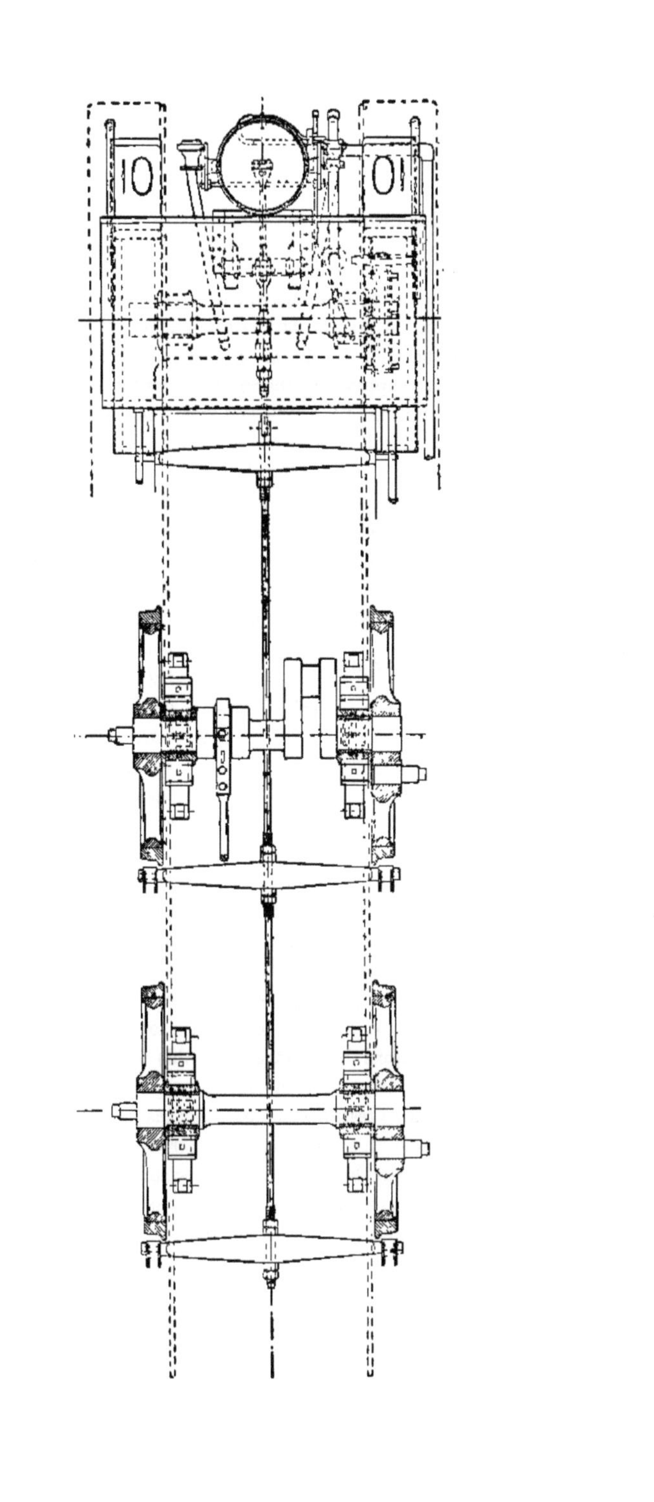

The four centres are now marked upon the face of the driving tire in the following manner, which gives the dead centres for each crank. A man is stationed in the pit to watch the progress of the slide blocks as the driving axle is turned. When they have approached the bump line within, say, 1 inch, he stops the turning of the axle, and from a convenient centre on the bottom slide, he scribes on the bar a mark of convenient length, between the centre of the block and the bump line. This position is also indicated upon the face of the tire. The axle is again turned slowly, and he notes as accurately as possible when the slide block stops, which gives the clearance between the cylinder end and piston. One leg of the dividers is then placed in the slide block centre and the axle turned until the point of the other leg covers the original scribed line, this position being also indicated upon the face of the tire, and by dividing the space between these two points on the face of the tire, the dead centre of the crank in that position is obtained accurately, and it does not depend upon the judgment of the operator in the least degree. This is repeated until all four centres are obtained. One crank is then placed at a dead centre, then the motion is reversed from full gear to full gear; if the valve spindle moves, the reversing shaft is either too high or low, and should be adjusted accordingly, reasons being given for this later on. Each crank should be tried on both centres, and at the same time the lead should be tested, which has been marked by trammel upon the spindle, this being equalised for the back and front port in both gears, by advancing or receding the reversing shaft according to requirements. This constitutes the whole mystery of setting the valves in the shop. It is simple

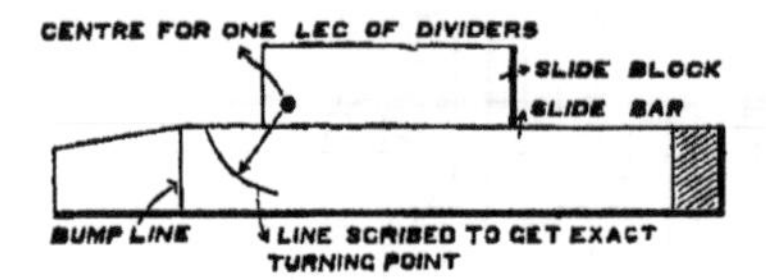

enough; but under the surface lie reasons which make it even more simple, certainly very much clearer, for knowing the reason of doing each operation essentially causes the whole to be done more scientifically and with greater accuracy. We must therefore turn to the drawing-office; but it is with a certain amount of diffidence, because the motion has now become so familiar, owing to its application to almost every existing class of engine, to its being illustrated in several of the engineering papers, and to its forming the subject for papers read before nearly every scientific body, that it appears almost superfluous to go very deep into detail; but at the same time, to show clearly the setting of the valves, this portion of the work would not be complete without it.

The essential conditions required to be known are the relative positions of the centre lines of the piston and valve spindle (the latter being in the plane of the vibration of the connecting-rod) the full stroke of the valve, also the stroke of the piston, the lap and the lead.

Fig. 309:—Lay down the centre line of the cylinder *a a*, and that of the valve spindle *b b*, at their relative distances. Draw the path of the crank pin, and the centre lines of the connecting-rod $c\ c^1$, $c\ c^{11}$, for both upper and lower positions, when the piston is at half stroke. Take a point *d*, on the centre line of the connecting-rod, where its vibration between d^1 and d^2 is equal to about double the length of the full stroke of the valve, allowing more rather than less, because it then renders a less angle of the reversing shaft when in full forward or backward gear. Through *d* draw the vertical *z z*, at right angles to *a a*, and mark off the two points $e\ e^1$, on each side, which are the extreme positions of the point *d*, for front and back strokes; from these points draw lines to a point *f* on the vertical, such that the angle between them shall not be more than 90°, less if there is room to allow it. Select a point f^1 either backward or forward, at a convenient

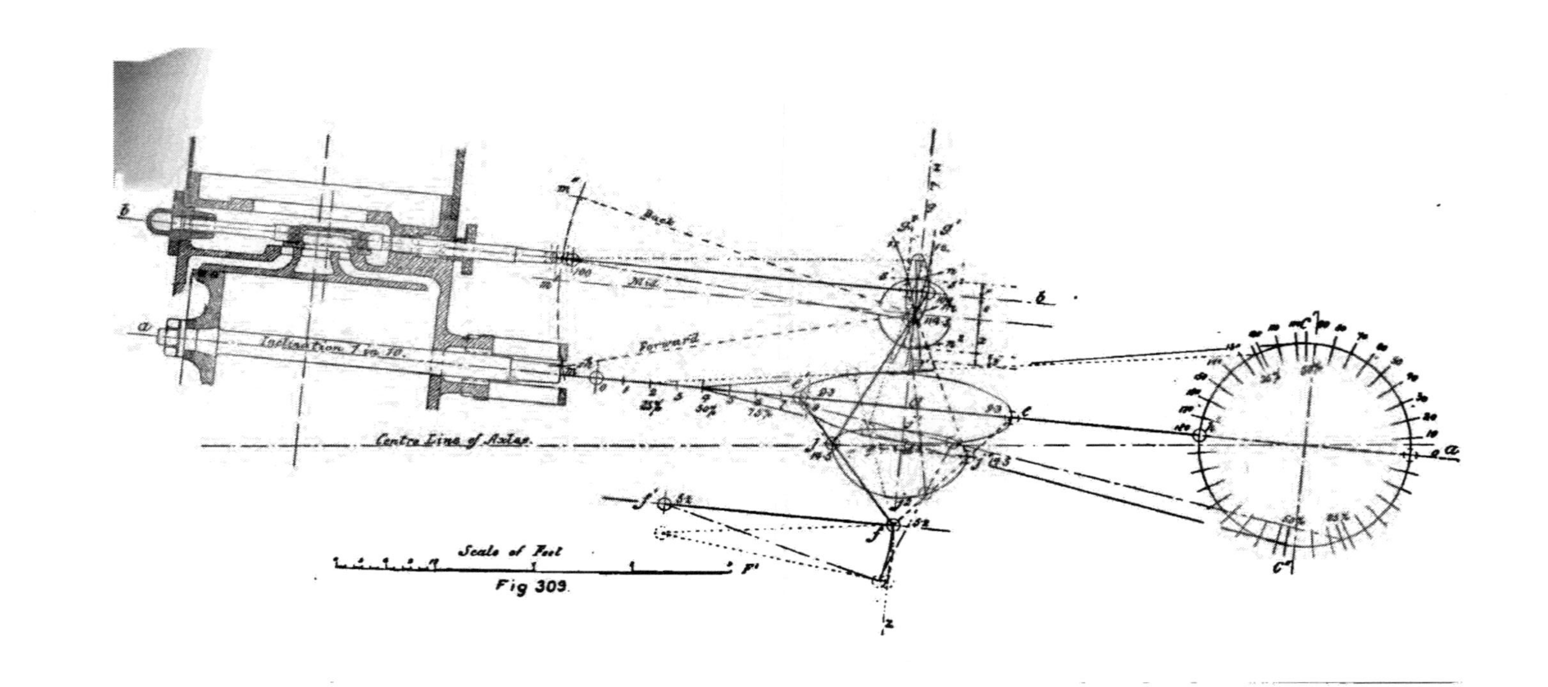
Back
Mid
Forward
Inclination 1 in 10.
Centre Line of Axles.
Scale of Feet

Fig 309

distance, so that the point f, which rises and falls with the connecting-rod, is maintained by the anchor link as nearly as possible on the vertical line. This vibration is shown in the figure to be all upon one side of the vertical, which does not make any material difference to the motion. The vertical is also indicated by the dotted lines.

Next, on the valve spindle centre line b b mark off upon each side of the vertical the amount required for lap and lead together, $g-g^1$ and $g-g^2$, the former being lap and lead for the front port and the latter for the back port. Assume the piston to be at the front end of the cylinder, then the point d will be at e^1, and the stirrup link at e^1 f. From a point on this link which has at first to be assumed, say at j—which will be about one-third more than the half vibration of the connecting-rod, that is d to d^1—draw the centre line of the lever actuating the valve, that is joining j and g^1; the point where this line crosses the vertical, z z, will be the centre of the fulcrum of the lever, and will also be the centre of oscillation of the curved links in the reversing shaft, in which the blocks carrying the centre of the lever slide. This centre is marked m, and stands for both centres, which must be concentric at each end of the stroke. From this construction, that is, by marking out the lap and the lead and actuating it by a lever, the steam is admitted at the commencement of each stroke by the amount of the lead only, and as this depends upon the link as a lever (which does not admit of a varying length) the lead remains constant for all degrees of expansion. Also because the point m was made the centre or fulcrum of the actuating lever and the centre of oscillation, which coincide when the piston is at each end of the stroke; when reversed from full gear to full gear, or any intermediate point, the valve should not move. This is the test always given to ascertain the correctness to which the valves have been set, nothing being admitted beyond a movement equal to the line of a fine scriber point.

It will be seen at a glance, and can be easily demonstrated with a pair of dividers, that if the point j of the lower end of the actuating lever were attached directly to the point e^1 on the connecting-rod, there would be imparted to the centre m, or fulcrum of that lever, an unequal vibration above and below the centre of the reversing shaft, which would give an unequal port and cut off for the two ends of the stroke. The function of the stirrup link $e^1 f$ and its attachment to the valve lever at j is to correct this error, for while the point e on the connecting-rod is performing a nearly true ellipse, the point j is moving in a figure like an ellipse bulged out at one side; but the axis $j^1 j^2$ is equal to the minor axis $d^1 d^2$, which is also equal to $n^1 n^2$. This irregularity is so set to correct the above error, and it gives an equal vibration to the centre m, above and below the centre of the reversing shaft. Although the position of the point j may be found by calculation, it is much more quickly found by a tentative process, and to test if the assumed point j be the correct one, we mark off on each side of m vertically the correct equal vibration required, $n^1 n^2$, which will be the same as $d^1 d^2$. Then apply the distance $e^1 j$ to $d^1 j^1$ and $d^2 j^2$. Then if the length $j n$ be applied to $j^1 n^1$ measuring from j^1, and to $j^2 n^2$ measuring from j^2, and the point m fall below $n^1 n^2$ in each case, it will be necessary to take a point on $e^1 f$ higher than j; or if, on the other hand, m fall above $n^1 n^2$, then the point j must be taken lower. This point will be very soon found, the only stipulation being that the length $j m$ of the lever $j m g^1$ must be such that its centre m vibrates equally on each side of the centre of the quadrant, also marked m.

The point g will be the point of attachment for the valve rod, which may be of any convenient length; but from that length as a radius, the centre of the links must be drawn from a centre m^1 on the parallel line m m^1, which corrects the error caused by the movement of the end of the valve rod. The angle at which this curve is set from the vertical —which is mid-gear—will give forward or backward gear, the angle leaning forward s^1 being forward gear and the reverse s^2 being backward gear, the centre of these curves being shown at m^2 and m^3. The amount of the angle, marked on the curve of extreme vibration at s s^1 or s s^2 will equal a quarter more than the full opening of the port at that angle, and the point of cut-off will be about 75 per cent. Thus if 1 inch opening of the port be required, the amount of the angle s s^1 must be 1¼ inch. Laid out in this form, the leads and cut-offs for both ends of the cylinder, and for backward and forward gear, will be practically perfect and equal, and the openings of the port nearly so. Deviations may be made without materially altering the correctness of the results.

During the passage of the point j the point m receives an acceleration due to the decreasing angularity of the rod up to the point j^1, and also during the remaining portion of the stroke owing to an increasing angularity. For the return stroke the point m encounters retardation for similar reasons. This is shown in Fig. 309, where R equals the amount due to the vertical movement of the point j, and P that due to the decreasing angle of the rod, whereas in the return stroke the whole of the movement of the point m would be represented by x, excepting for the retardation represented by the minus quantity y. This peculiar movement occurs just at the right moment. The valve opens and admits steam in ample volume at the commencement, and then pauses when full open, due to this peculiarity. It also opens to exhaust promptly to the full extent, then pauses, so that the steam escapes in the same free manner that it was admitted. These advantages can be more forcibly realised by comparing the valve path diagrams of sister engines, one having this gear and the other the link motion, both having the same amount of lap and cut-off. Further advantages are:—Every portion is always in useful action, cylinders can be placed nearer together and increased bearings attained, it is accessible and easy to reverse, besides having less area by one-quarter exposed to friction than the link motion.

In Fig. 309, if 100 represents the total amount of resistance to the progress of the valve, then the different values of the stress upon the other pins are represented accordingly.

Fig. 308:—Now that the valves have been set, the holes may be opened out in the frames for the reversing shaft brackets and four 1-inch bolts driven home into each. In the meantime the foot plate end has been receiving attention, for the cab has been fitted, put on, and fastened to the platform with bolts. The cab sides and splashers are formed of one plate outside, the cast-iron sand-box being fitted to this plate inside, the reversing bracket being bolted to the top of the left trailing splasher. Now that the two sides are fixed, it only remains for the front and top to be placed, when the cab will be complete. The vacuum brake cylinder is then placed in position by securing it to the underside of the foot plate by carrier plates or brackets. Then all vacuum and injector pipes, &c. are coupled up, also the handrail round the boiler and smoke-box front is secured to its eye bolts or studs. The brake hangers are then hung on their studs, the cross bars fixed, and all rods coupled up. This completes the work as far as shown in the figures; what remains to be done, which is about six or seven hours' work, is shown on the sectional drawing. This consists of securing the buffer-plate, buffers, draw-bars, fall plate, foot steps, packing glands, &c. The engine is then slung in the cranes and the coupling rods put on, the springs adjusted, carried to the traverser, and then drawn out of the shop. The tender is then coupled to the engine, the brake blocks adjusted, also all necessary pipes, and then fired up ready for trial, which consists of about twenty to twenty-five miles run.

During the trial, observations should be made relative to the clearance spaces in and about the motion and elsewhere, also that all pins, journals and bearings remain cool. The injectors should be perfectly free and easy to work. The regulator, reversing, sand, cylinder cock gear and water pick-up should work freely, but without any undue slackness, this also being applicable to all other details. The vacuum brake must receive special attention, it being thoroughly tested as to leakage, amount of vacuum and power. The latter is generally tried by putting the brake full on and opening the regulator full. All the joints on the foot plate for boiler mountings, the dome and safety valve covers, also the cylinder and steam chest covers, should be perfectly tight. Finally, each wheel should be run over a $\frac{1}{4}$ or $\frac{3}{8}$-inch washer, and the effect upon the springs noted, which should give accordingly. The engine is then handed over to the paint shop.

Here it first receives a thorough scouring all over with sandstone, and is afterwards washed down with turpentine, to thoroughly cleanse it from all rust and dirt. It is then given one coat of oil lead colour, which consists of white lead and common black, mixed with boiled linseed oil, turpentine and terebene drier. This coat gives adherence to the stopping and filling, which consists of white lead, Indian copal varnish and gold size. The whole surface of the engine is gone over first, and the worst parts filled up with a thick stopping, using putty knives, and then followed with a thinner stopping worked on with trowels. The rivets are then brushed round with a thinner filling, which softens that put on with trowels, and makes the whole a smooth surface. A cheaper material is mixed with the white lead and used after the first coat of lead colour, when the surfaces are worse than usual and a great quantity of material is required. It is then stained with one coat of vegetable black, mixed with gold size and turpentine, which acts as a guide for the rubbers-down. A smooth surface is then got up by wet rubbing with Schumachersche's Fabrick. Afterwards it receives the first coat of paint, which is a dark lead, mixed in a similar manner to the light lead colour used before filling up. This is followed by a coat of the best drop ivory black, which is mixed with gold size and turpentine, bound with varnish. The third coat consists of the best drop ivory black mixed with varnish, upon which the lining out is done. It is then ready for the varnish, the first two coats being flattened down with pumice powder, horsehair and water, followed by a third coat; best engine copal varnish being used in all cases. The cab is filled up inside in a similar manner to the rest of the engine, and painted with three coats of buff or stone colour, which consists of white lead, Turkey burnt umber, orange chrome, mixed with boiled oil, turpentine and terebene driers. It is then stencilled, lined out, and given two coats of clear varnish. All the motion work, where not bright, and the buffer beam, receives three coats of vermilion and varnished. Wheels, framing, smoke-box and brake gear receive one coat of drop ivory black and two coats of the best Japan black. The whole operation occupies about three weeks, including one week for the varnish to set, and is such that it will not be required to be repeated for five years.

INDEX.

LONDON: PRINTED BY WILLIAM CLOWES AND SONS, LIMITED, STAMFORD STREET AND CHARING CROSS.

BOOKS RELATING

TO

APPLIED SCIENCE

PUBLISHED BY

E. & F. N. SPON.

LONDON: 125 STRAND.

NEW YORK: 12 CORTLANDT STREET.

Algebra.—*Algebra Self-Taught.* By W. P. HIGGS, M.A., D.Sc., LL.D., Assoc. Inst. C.E., Author of 'A Handbook of the Differential Calculus,' etc. Second edition, crown 8vo, cloth, 2*s.* 6*d.*

CONTENTS:

Symbols and the Signs of Operation—The Equation and the Unknown Quantity—Positive and Negative Quantities—Multiplication—Involution—Exponents—Negative Exponents—Roots, and the Use of Exponents as Logarithms—Logarithms—Tables of Logarithms and Proportionate Parts—Transformation of System of Logarithms—Common Uses of Common Logarithms—Compound Multiplication and the Binomial Theorem—Division, Fractions, and Ratio—Continued Proportion—The Series and the Summation of the Series—Limit of Series—Square and Cube Roots—Equations—List of Formulæ, etc.

Architects' Handbook.—*A Handbook of Formulæ, Tables and Memoranda, for Architectural Surveyors and others engaged in Building.* By J. T. HURST, C.E. Fourteenth edition, royal 32mo, roan, 5*s.*

"It is no disparagement to the many excellent publications we refer to, to say that in our opinion this little pocket-book of Hurst's is the very best of them all, without any exception. It would be useless to attempt a recapitulation of the contents, for it appears to contain almost *everything* that anyone connected with building could require, and, best of all, made up in a compact form for carrying in the pocket, measuring only 5 in. by 3 in., and about ¾ in. thick, in a limp cover. We congratulate the author on the success of his laborious and practically compiled little book, which has received unqualified and deserved praise from every professional person to whom we have shown it."—*The Dublin Builder.*

Architecture.—*Town and Country Mansions and Suburban Houses*, with Notes on the Sanitary and Artistic Construction of Houses, *illustrated by* 30 *plates*, containing Plans, Elevations, Perspectives, and Interior Views of Executed Works in the Queen Anne, Classic, Old English, Adams, Jacobean, Louis XVI., and other Styles. By WILLIAM YOUNG, Author of 'Picturesque Architectural Studies.' Imp. 4to, cloth, 10*s.* 6*d.*

1894.

RELATING
O

SCIENCE
D BY

SPON.

STRAND.

ANDT STREET.

ht. By W. P. HIGGS,
uthor of 'A Handbook of the
, crown 8vo, cloth, 2s. 6d.

n and the Unknown Quantity—
ion—Exponents—Negative Expo-
ogarithms—Tables of Logarithms
Logarithms—Common Uses of
Binomial Theorem—Division,
l the Summation of the Series—
f Formulæ, etc.

andbook of For-
il Surveyors and others
ourteenth edition, royal

e refer to, to say that in our
all, without any exception,
it appears to contain almost
d, best of all, made up in a
in., and about ½ in. thick,
laborious and practically
praise from every pro

n,
517
rav-
6d.
nce of
—Sec-
titions—
entres for
n Bridges
Timber.

Mansions and
tistic Constr
Elevation
he Quee
nd othe
ectural

emistry.
at the
bury,
I.C.

Architecture.—*The Seven Periods of English Architecture*, defined and illustrated. By EDMUND SHARPE, M.A., Architect. 20 *steel engravings and 7 woodcuts*, third edition, royal 8vo, cloth, 12*s*. 6*d*.

Assaying.—*The Assayer's Manual*: an Abridged Treatise on the Docimastic Examination of Ores and Furnace and other Artificial Products. By BRUNO KERL. Translated by W. T. BRANNT. *With* 65 *illustrations*, 8vo, cloth, 12*s*. 6*d*.

Baths.—*The Turkish Bath*: its Design and Construction for Public and Commercial Purposes. By R. O. ALLSOP, Architect. *With plans and sections*, 8vo, cloth, 6*s*.

Baths and Wash Houses.—*Public Baths and Wash Houses*. By ROBERT OWEN ALLSOP, Architect, Author of 'The Turkish Bath,' &c. *With cuts and folding plates*, demy 8vo, cloth, 6*s*.

Blasting.—*Rock Blasting*: a Practical Treatise on the means employed in Blasting Rocks for Industrial Purposes. By G. G. ANDRÉ, F.G.S., Assoc. Inst. C.E. *With* 56 *illustrations and* 12 *plates*, 8vo, cloth, 5*s*.

Boilers.—*A Pocket-Book for Boiler Makers and Steam Users*, comprising a variety of useful information for Employer and Workman, Government Inspectors, Board of Trade Surveyors, Engineers in charge of Works and Slips, Foremen of Manufactories, and the general Steam-using Public. By MAURICE JOHN SEXTON. Third edition, enlarged, royal 32mo, roan, gilt edges, 5*s*.

Boilers.—*The Boiler-Maker's & Iron Ship-Builder's Companion*, comprising a series of original and carefully calculated tables, of the utmost utility to persons interested in the iron trades. By JAMES FODEN, author of 'Mechanical Tables,' etc. Second edition, revised, *with illustrations*, crown 8vo, cloth, 5*s*.

Brass Founding.—*The Practical Brass and Iron-Founder's Guide*, a Treatise on the Art of Brass Founding, Moulding, the Metals and their Alloys, etc. By JAMES LARKIN. New edition, revised and greatly enlarged, crown 8vo, cloth, 10*s*. 6*d*.

Breweries.—*Breweries and Maltings*: their Arrangement, Construction, Machinery, and Plant. By G. SCAMELL, F.R.I.B.A. Second edition, revised, enlarged, and partly rewritten. By F. COLYER, M.I.C.E., M.I.M.E. *With* 20 *plates*, 8vo, cloth, 12*s*. 6*d*.

Brewing.—*A Text Book of the Science of Brewing*. BY EDWARD RALPH MORITZ, Chemist to the Country Brewers' Society, and GEORGE HARRIS MORRIS, Ph.D., F.C.S., F.I.C., etc. Based upon a course of six lectures delivered by E. R. MORITZ at the Finsbury Technical College of the City and Guilds of London Institute. *With plates and illustrations*, 8vo, cloth, 1*l*. 1*s*.

Bridges.—*Elementary Theory and Calculation of Iron Bridges and Roofs.* By AUGUST RITTER, Ph.D., Professor at the Polytechnic School at Aix-la-Chapelle. Translated from the third German edition, by H. R. SANKEY, Capt. R.E. *With 500 illustrations*, 8vo, cloth, 15*s.*

Bridges.—*Stresses in Girder and Roof Trusses for both Dead and Live Loads by Simple Multiplication*, with Stress Constants for 100 cases, for the use of Civil and Mechanical Engineers, Architects and Draughtsmen. By F. R. JOHNSON, Assoc. M. Inst. C.E. Part 1, Girders. Part 2, Roofs. In 1 vol., crown 8vo, cloth, 6*s.*

CONTENTS:

Part 1.—Introductory. Part 2.—Stress Constants for Dead and Live Loads. Part 3.—Stress Diagrams.

Builders' Price Book.—*Spons' Architects' and Builders' Price Book, with useful Memoranda.* By W. YOUNG. Crown 8vo, cloth, red edges, 3*s.* 6*d.* *Published annually.*

Building.—*The Clerk of Works:* a Vade-Mecum for all engaged in the Superintendence of Building Operations. By G. G. HOSKINS, F.R.I.B.A. Third edition, fcap. 8vo, cloth, 1*s.* 6*d.*

Building.—*The Builder's Clerk:* a Guide to the Management of a Builder's Business. By THOMAS BALES. Fcap. 8vo, cloth, 1*s.* 6*d.*

Canals.—*Waterways and Water Transport in Different Countries.* With a description of the Panama, Suez, Manchester, Nicaraguan, and other Canals. By J. STEPHEN JEANS, Author of 'England's Supremacy,' 'Railway Problems,' &c. *Numerous illustrations*, 8vo, cloth, 14*s.*

Carpentry.—*The Elementary Principles of Carpentry.* By THOMAS TREDGOLD. Revised from the original edition, and partly re-written, by JOHN THOMAS HURST. Contained in 517 pages of letterpress, and *illustrated with 48 plates and 150 wood engravings.* Sixth edition, reprinted from the third, crown 8vo, cloth, 12*s.* 6*d.*

Section I. On the Equality and Distribution of Forces—Section II. Resistance of Timber—Section III. Construction of Floors—Section IV. Construction of Roofs—Section V. Construction of Domes and Cupolas—Section VI. Construction of Partitions—Section VII. Scaffolds, Staging, and Gantries—Section VIII. Construction of Centres for Bridges—Section IX. Coffer-dams, Shoring, and Strutting—Section X. Wooden Bridges and Viaducts—Section XI. Joints, Straps, and other Fastenings—Section XII. Timber.

Chemistry.—*Practical Work in Organic Chemistry.* By F. W. STREATFEILD, F.I.C., etc., Demonstrator of Chemistry at the City and Guilds of London Institutes Technical College, Finsbury. With a Prefatory Notice by Professor R. MELDOLA, F.R.S., F.I.C. Crown 8vo, cloth, 3*s.*

Chemists' Pocket Book.—*A Pocket-Book for Chemists, Chemical Manufacturers, Metallurgists, Dyers, Distillers, Brewers, Sugar Refiners, Photographers, Students, etc., etc.* By THOMAS BAYLEY, Assoc. R.C. Sc. Ireland, Analytical and Consulting Chemist and Assayer. Fifth edition, 481 pp., royal 32mo, roan, gilt edges, 5*s.*

SYNOPSIS OF CONTENTS:

Atomic Weights and Factors—Useful Data—Chemical Calculations—Rules for Indirect Analysis—Weights and Measures—Thermometers and Barometers—Chemical Physics—Boiling Points, etc.—Solubility of Substances—Methods of Obtaining Specific Gravity—Conversion of Hydrometers—Strength of Solutions by Specific Gravity—Analysis—Gas Analysis—Water Analysis—Qualitative Analysis and Reactions—Volumetric Analysis—Manipulation—Mineralogy—Assaying—Alcohol—Beer—Sugar—Miscellaneous Technological matter relating to Potash, Soda, Sulphuric Acid, Chlorine, Tar Products, Petroleum, Milk, Tallow, Photography, Prices, Wages, Appendix, etc., etc.

Coal Mining.—*A Glossary of Terms used in Coal Mining.* By WILLIAM STUKELEY GRESLEY, Assoc. Mem. Inst. C.E., F.G.S., Member of the North of England Institute of Mining Engineers. *Illustrated with numerous woodcuts and diagrams*, crown 8vo, cloth, 5*s.*

Coffee Cultivation.—*Coffee: its Culture and Commerce in all Countries.* Edited by C. G. WARNFORD LOCK, F.L.S. Crown 8vo, cloth, 12*s.* 6*d.*

A practical handbook for the Planter, treating in a thoroughly practical manner on the cultivation of the Plant, the management of an estate, Diseases and enemies of the Coffee Plant (with their prevention and cure), preparation of the berry for market, and statistics of local details of culture and production. Bibliography.

Colonial Engineering.—*Spons' Information for Colonial Engineers.* Edited by J. T. HURST. Demy 8vo, sewed.

No. 1, Ceylon. By ABRAHAM DEANE, C.E. 2*s.* 6*d.*

Introductory Remarks—Natural Productions—Architecture and Engineering—Topography, Trade, and Natural History—Principal Stations—Weights and Measures, etc., etc.

No. 2. Southern Africa, including the Cape Colony, Natal, and the Dutch Republics. By HENRY HALL, F.R.G.S., F.R.C.I. With Map. 3*s.* 6*d.*

General Description of South Africa—Physical Geography with reference to Engineering Operations—Notes on Labour and Material in Cape Colony—Geological Notes on Rock Formation in South Africa—Engineering Instruments for Use in South Africa—Principal Public Works in Cape Colony: Railways, Mountain Roads and Passes, Harbour Works, Bridges, Gas Works, Irrigation and Water Supply, Lighthouses, Drainage and Sanitary Engineering, Public Buildings, Mines—Table of Woods in South Africa—Animals used for Draught Purposes—Statistical Notes—Table of Distances—Rates of Carriage, etc.

No. 3. India. By F. C. DANVERS, Assoc. Inst. C.E. With Map. 4*s.* 6*d.*

Physical Geography of India—Building Materials—Roads—Railways—Bridges—Irrigation—River Works—Harbours—Lighthouse Buildings—Native Labour—The Principa Trees of India—Money—Weights and Measures—Glossary of Indian Terms, etc.

Concrete.—*Notes on Concrete and Works in Concrete;* especially written to assist those engaged upon Public Works. By JOHN NEWMAN, Assoc. Mem. Inst. C.E. Second edition, revised and enlarged, crown 8vo, cloth, 6*s.*

Depreciation of Factories.—*The Depreciation of Factories and their Valuation.* By EWING MATHESON, Mem. Inst. C.E. Second edition, revised, with an Introduction by W. C. JACKSON, Member of the Council of the Institute of Chartered Accountants. 8vo, cloth, 7*s.* 6*d.*

Drainage.—*The Drainage of Fens and Low Lands by Gravitation and Steam Power.* By W. H. WHEELER, M. Inst. C.E. *With plates,* 8vo, cloth, 12*s.* 6*d.*

Drawing,—*Hints on Architectural Draughtsmanship.* By G. W. TUXFORD HALLATT. Fcap. 8vo, cloth, 1*s.* 6*d.*

Drawing.—*The Draughtsman's Handbook of Plan and Map Drawing;* including instructions for the preparation of Engineering, Architectural, and Mechanical Drawings. *With numerous illustrations in the text, and* 33 *plates* (15 *printed in colours*). By G. G. ANDRÉ, F.G.S., Assoc. Inst. C.E. 4to, cloth, 9*s.*

Drawing Instruments.—*A Descriptive Treatise on Mathematical Drawing Instruments:* their construction, uses, qualities, selection, preservation, and suggestions for improvements, with hints upon Drawing and Colouring. By W. F. STANLEY, M.R.I. Sixth edition, *with numerous illustrations,* crown 8vo, cloth, 5*s.*

Dynamo.—*Dynamo-Tenders' Hand-Book.* By F. B. BADT. *With* 70 *illustrations.* Third edition, 18mo, cloth, 4*s.* 6*d.*

Dynamo.—*Theoretical Elements of Electro-Dynamic Machinery.* By A. E. KENNELLY. *With illustrations,* 8vo, cloth, 4*s.* 6*d.*

Dynamo-Electric Machinery.—*Dynamo-Electric Machinery:* a Text-Book for Students of Electro-Technology. By SILVANUS P. THOMPSON, B.A., D.Sc. 8vo, cloth.

Earthwork Slips.—*Earthwork Slips and Subsidences upon Public Works:* Their Causes, Prevention and Reparation. Especially written to assist those engaged in the Construction or Maintenance of Railways, Docks, Canals, Waterworks, River Banks, Reclamation Embankments, Drainage Works, &c., &c. By JOHN NEWMAN, Assoc. Mem. Inst. C.E., Author of 'Notes on Concrete,' &c. Crown 8vo, cloth, 7*s.* 6*d.*

Electric Bells.—*Electric Bell Construction:* a treatise on the construction of Electric Bells, Indicators, and similar apparatus. By F. C. ALLSOP, Author of 'Practical Electric Bell Fitting.' *With* 177 *illustrations drawn to scale,* crown 8vo, cloth, 3*s.* 6*d.*

Electric Bells.—*A Practical Treatise on the fitting up and maintenance of Electric Bells and all the necessary apparatus.* By F. C. ALLSOP, Author of 'Telephones, their Construction and Fitting.' Second edition, revised, *nearly* 150 *illustrations*, crown 8vo, cloth, 3*s.* 6*d.*

Electric Lighting.—*Wrinkles in Electric Lighting.* By VINCENT STEPHEN. *With illustrations.* 18mo, cloth, 2*s.* 6*d.*

CONTENTS:

1. The Electric Current and its production by Chemical means—2. Production of Electric Currents by Mechanical means—3. Dynamo-Electric Machines—4. Electric Lamps—5. Lead—6. Ship Lighting.

Electric Telegraph. — *Telegraphic Connections,* embracing recent methods in Quadruplex Telegraphy. By CHARLES THOM and WILLIS H. JONES. *With illustrations.* Oblong 8vo, cloth, 7*s.* 6*d.*

Electric Testing.—*A Guide for the Electric Testing of Telegraph Cables.* By Col. V. HOSKICER, Royal Danish Engineers. Third edition, crown 8vo, cloth, 4*s.* 6*d.*

Electric Telegraph.—*A History of Electric Telegraphy,* to the Year 1837. Chiefly compiled from Original Sources and hitherto Unpublished Documents, by J. J. FAHIE, Mem. Soc. of Tel. Engineers, and of the International Society of Electricians, Paris. Crown 8vo, cloth, 9*s.*

Electric Toys.—*Electric Toys.* Electric Toy-Making, Dynamo Building and Electric Motor Construction for Amateurs. By T. O'CONOR SLOANE, Ph.D. *With cuts,* crown 8vo, cloth, 4*s.* 6*d.*

Electrical Notes.—*Practical Electrical Notes and Definitions for the use of Engineering Students and Practical Men.* By W. PERREN MAYCOCK, Assoc. M. Inst. E.E., Instructor in Electrical Engineering at the Pitlake Institute, Croydon, together with the Rules and Regulations to be observed in Electrical Installation Work. Second edition. Royal 32mo, cloth, red edges, 3*s.*

Electrical Tables.—*Electrical Tables and Memoranda.* By SILVANUS P. THOMPSON, D.Sc., B.A., F.R.S., and EUSTACE THOMAS. In waistcoat-pocket size (2½ in. by 1⅝ in.), French morocco, gilt edges, *with numerous illustrations,* 1*s.*

Electrical Testing.—*A Handbook of Electrical Testing.* By H. R. KEMPE, M.I.E.E. Fourth edition, revised and enlarged, 8vo, cloth, 18*s.*

Electrical Testing.—*A Practical Guide to the Testing of Insulated Wires and Cables.* By HERBERT LAWS WEBB, Member of the American Institute of Electrical Engineers, and of the Institution of Electrical Engineers, London. Crown 8vo, cloth, 4*s.* 6*d.*

Electricity.—*The Arithmetic of Electricity:* a Manual of Electrical Calculations by Electrical Methods. By T. O'CONOR SLOANE. Crown 8vo, cloth, 4*s.* 6*d.*

Electricity.—*Short Lectures to Electrical Artisans,* being a Course of Experimental Lectures delivered to a practical audience. By J. A. FLEMING, M.A., D.Sc. (Lond.), Professor of Electrical Technology in University College, London. *With diagrams,* fourth edition, crown 8vo, cloth, 4*s.*

Electricity.—*Electricity, its Theory, Sources, and Applications.* By JOHN T. SPRAGUE, M. Inst. E.E. Third edition, thoroughly revised and extended, *with numerous illustrations and tables,* crown 8vo, cloth, 15*s.*

Electricity.—*Transformers: their Theory, Construction, and Application Simplified.* By C. D. HASKINS, Assoc. Mem. American Institute of Electrical Engineers. *Illustrated,* crown 8vo, cloth, 4*s.* 6*d.*

Electricity in the House.—*Domestic Electricity for Amateurs.* Translated from the French of E. HOSPITALIER, Editor of 'L'Electricien,' by C. J. WHARTON, M. Inst. E.E. *Numerous illustrations.* Demy 8vo, cloth, 6*s.*

CONTENTS:

1. Production of the Electric Current—2. Electric Bells—3. Automatic Alarms—4. Domestic Telephones—5. Electric Clocks—6. Electric Lighters—7. Domestic Electric Lighting—8. Domestic Application of the Electric Light—9. Electric Motors—10. Electrical Locomotion—11. Electrotyping, Plating, and Gilding—12. Electric Recreations—13. Various applications—Workshop of the Electrician.

Electro-Magnet.—*The Electro-Magnet and Electromagnetic Mechanism.* By SILVANUS P. THOMPSON, D.Sc., F.R.S. *With 213 illustrations.* Second edition, 8vo, cloth, 15*s.*

Electro-Motors.—*Notes on design of Small Dynamo.* By GEO. HALLIDAY, Whitworth Scholar, Professor of Engineering at the Hartley Institute, Southampton. *Plates,* 8vo, cloth, 2*s.* 6*d.*

Electro-Motors.—*The practical management of Dynamos and Motors.* By FRANCIS B. CROCKER, Professor of Electrical Engineering, Columbia College, New York, and SCHUYLER S. WHEELER, D.Sc. *Cuts,* crown 8vo, cloth, 4*s.* 6*d.*

Engineering Drawing. — *Practical Geometry, Perspective and Engineering Drawing*; a Course of Descriptive Geometry adapted to the Requirements of the Engineering Draughtsman, including the determination of cast shadows and Isometric Projection, each chapter being followed by numerous examples; to which are added rules for Shading, Shade-lining, etc., together with practical instructions as to the Lining, Colouring, Printing, and general treatment of Engineering Drawings, with a chapter on drawing Instruments. By GEORGE S. CLARKE, Capt. R.E. Second edition, *with* 21 *plates*. 2 vols., cloth, 10*s*. 6*d*.

Engineers' Tables.—*A Pocket-Book of Useful Formulæ and Memoranda for Civil and Mechanical Engineers*. By Sir GUILFORD L. MOLESWORTH, Mem. Inst. C.E., and R. B. MOLESWORTH. *With numerous illustrations*, 782 pp. Twenty-third edition, 32mo, roan, 6*s*.

SYNOPSIS OF CONTENTS:

Surveying, Levelling, etc.—Strength and Weight of Materials—Earthwork, Brickwork, Masonry, Arches, etc.—Struts, Columns, Beams, and Trusses—Flooring, Roofing, and Roof Trusses—Girders, Bridges, etc.—Railways and Roads—Hydraulic Formulæ—Canals, Sewers, Waterworks, Docks—Irrigation and Breakwaters—Gas, Ventilation, and Warming—Heat, Light, Colour, and Sound—Gravity: Centres, Forces, and Powers—Millwork, Teeth of Wheels, Shafting, etc.—Workshop Recipes—Sundry Machinery—Animal Power—Steam and the Steam Engine—Water-power, Water-wheels, Turbines, etc.—Wind and Windmills—Steam Navigation, Ship Building, Tonnage, etc.—Gunnery, Projectiles, etc.—Weights, Measures, and Money—Trigonometry, Conic Sections, and Curves—Telegraphy—Mensuration—Tables of Areas and Circumference, and Arcs of Circles—Logarithms, Square and Cube Roots, Powers—Reciprocals, etc.—Useful Numbers—Differential and Integral Calculus—Algebraic Signs—Telegraphic Construction and Formulæ.

Engineers' Tables.—*Spons' Tables and Memoranda for Engineers*. By J. T. HURST, C.E. Twelfth edition, revised and considerably enlarged, in waistcoat-pocket size ($2\frac{5}{8}$ in. by 2 in.), roan, gilt edges, 1*s*.

Experimental Science.—*Experimental Science*: Elementary, Practical, and Experimental Physics. By GEO. M. HOPKINS. *Illustrated by* 890 *engravings*. 840 pp., 8vo, cloth, 16*s*.

Factories.—*Our Factories, Workshops, and Warehouses*: their Sanitary and Fire-Resisting Arrangements. By B. H. THWAITE, Assoc. Mem. Inst. C.E. *With* 183 *wood engravings*, crown 8vo, cloth, 9*s*.

Foundations.—*Notes on Cylinder Bridge Piers and the Well System of Foundations*. By JOHN NEWMAN, Assoc. M. Inst. C.E., 8vo, cloth, 6*s*.

Founding.—*A Practical Treatise on Casting and Founding*, including descriptions of the modern machinery employed in the art. By N. E. SPRETSON, Engineer. Fifth edition, with 82 *plates* drawn to scale, 412 pp., demy 8vo, cloth, 18*s*.

Founding.—*American Foundry Practice:* Treating of Loam, Dry Sand, and Green Sand Moulding, and containing a Practical Treatise upon the Management of Cupolas, and the Melting of Iron. By T. D. WEST, Practical Iron Moulder and Foundry Foreman. Second edition, *with numerous illustrations*, crown 8vo, cloth, 12*s*. 6*d*.

French Polishing. — *The French - Polisher's* Manual. By a French-Polisher; containing Timber Staining, Washing, Matching, Improving, Painting, Imitations, Directions for Staining, Sizing, Embodying, Smoothing, Spirit Varnishing, French-Polishing, Directions for Repolishing. Third edition, royal 32mo, sewed, 6*d*.

Furnaces.—*Practical Hints on the Working and Construction of Regenerator Furnaces*, being an Explanatory Treatise on the System of Gaseous Firing applicable to Retort Settings in Gas Works. By MAURICE GRAHAM, Assoc. Mem. Inst. C.E. *Cuts*, 8vo, cloth.

Gas Analysis.—*The Gas Engineers' Laboratory Handbook.* By JOHN HORNBY, F.I.C., Honours Medallist in Gas Manipulation, City and Guilds of London Institute. *Numerous illustrations*, crown 8vo, cloth, 6*s*.

CONTENTS:

The Balance—Weights and Weighing—Sampling—Mechanical Division—Drying and Desiccation — Solution and Evaporation — Precipitation — Filtration and Treatment of Precipitates — Simple Gravimetric Estimations — Volumetric Analyses—Special Analyses required by Gas Works—Technical Gas Analysis—Gas Referees' Instructions, etc. etc.

Gas Engines.—*Gas and Petroleum Engines:* a Practical Treatise on the Internal Combustion Engine. By WM. ROBINSON, M.E., Senior Demonstrator and Lecturer on Applied Mechanics, Physics, &c., City and Guilds of London College, Finsbury, Assoc. Mem. Inst. C.E., &c. *Numerous illustrations.* 8vo, cloth, 14*s*.

Gas Engineering.—*Manual for Gas Engineering Students.* By D. LEE. 18mo, cloth, 1*s*.

Gas Works.—*Gas Works:* their Arrangement, Construction, Plant, and Machinery. By F. COLYER, M. Inst. C.E. *With* 31 *folding plates*, 8vo, cloth, 12*s*. 6*d*.

Gold Mining.—*Practical Gold-Mining:* a Comprehensive Treatise on the Origin and Occurrence of Gold-bearing Gravels, Rocks and Ores, and the Methods by which the Gold is extracted. By C. G. WARNFORD LOCK, co-Author of 'Gold: its Occurrence and Extraction.' *With* 8 *plates and* 275 *engravings in the text*, 788 pp., royal 8vo, cloth, 2*l*. 2*s*.

Graphic Statics.—*The Elements of Graphic Statics.* By Professor KARL VON OTT, translated from the German by G. S. CLARKE, Capt. R.E., Instructor in Mechanical Drawing, Royal Indian Engineering College. *With* 93 *illustrations*, crown 8vo, cloth, 5*s*.

Graphic Statics.—*The Principles of Graphic Statics.* By GEORGE SYDENHAM CLARKE, Capt. Royal Engineers. *With* 112 *illustrations.* Second edition, 4to, cloth, 12*s.* 6*d.*

Graphic Statics.—*Mechanical Graphics.* A Second Course of Mechanical Drawing. With Preface by Prof. PERRY, B.Sc., F.R.S. Arranged for use in Technical and Science and Art Institutes, Schools and Colleges, by GEORGE HALLIDAY, Whitworth Scholar. *With illustrations,* 8vo, cloth, 6*s.*

Graphic Statics.—*A New Method of Graphic Statics,* applied in the construction of Wrought-Iron Girders, practically illustrated by a series of Working Drawings of modern type. By EDMUND OLANDER, of the Great Western Railway, Assoc. Mem. Inst. C.E. Small folio, cloth, 10*s.* 6*d.*

Heat Engine.—*Theory and Construction of a Natural Heat Motor.* Translated from the German of RUDOLF DIESEL by BRYAN DONKIN, Mem. Inst. C.E. *Numerous cuts and plates,* 8vo, cloth, 6*s.*

Hot Water.—*Hot Water Supply:* a Practical Treatise upon the Fitting of Circulating Apparatus in connection with Kitchen Range and other Boilers, to supply Hot Water for Domestic and General Purposes. With a Chapter upon Estimating. By F. DYE. *With illustrations,* crown 8vo, cloth, 3*s.*

Hot Water.—*Hot Water Apparatus:* an Elementary Guide for the Fitting and Fixing of Boilers and Apparatus for the Circulation of Hot Water for Heating and for Domestic Supply, and containing a Chapter upon Boilers and Fittings for Steam Cooking. By F. DYE. 32 *illustrations,* fcap. 8vo, cloth, 1*s.* 6*d.*

Household Manual.—*Spons' Household Manual:* a Treasury of Domestic Receipts and Guide for Home Management. Demy 8vo, cloth, containing 975 pages and 250 *illustrations,* price 7*s.* 6*d.*

PRINCIPAL CONTENTS:

Hints for selecting a good House—Sanitation—Water Supply—Ventilation and Warming—Lighting—Furniture and Decoration—Thieves and Fire—The Larder—Curing Foods for lengthened Preservation—The Dairy—The Cellar—The Pantry—The Kitchen—Receipts for Dishes—The Housewife's Room—Housekeeping, Marketing—The Dining-Room—The Drawing-Room—The Bedroom—The Nursery—The Sick-Room—The Bath-Room—The Laundry—The School-Room—The Playground—The Work-Room—The Library—The Garden—The Farmyard—Small Motors—Household Law.

House Hunting.—*Practical Hints on Taking a House.* By H. PERCY BOULNOIS, Mem. Inst. C.E., City Engineer, Liverpool, Author of 'The Municipal and Sanitary Engineer's Handbook,' 'Dirty Dustbins and Sloppy Streets,' &c. 18mo, cloth, 1*s.* 6*d.*

Hydraulics.—*Simple Hydraulic Formulæ.* By T. W. STONE, C.E., late Resident District Engineer, Victoria Water Supply. Crown 8vo, cloth, 4*s.*

Hydraulic Machinery.—*Hydraulic Steam and Hand-Power Lifting and Pressing Machinery.* By FREDERICK COLYER, M. Inst. C.E., M. Inst. M.E. Second edition, revised and enlarged. *With 88 plates,* 8vo, cloth, 28*s.*

Hydropathic Establishments.—*The Hydropathic Establishment and its Baths.* By R. O. ALLSOP, Architect. Author of 'The Turkish Bath.' *Illustrated with plates and sections,* 8vo, cloth, 5*s.*

CONTENTS:

General Considerations—Requirements of the Hydropathic Establishment—Some existing Institutions—Baths and Treatments and the arrangement of the Bath-House—Vapour Baths and the Russian Bath—The Douche Room and its appliances—Massage and Electrical Treatment—Pulverisation and the Mont Dore Cure—Inhalation and the Pine Cure—The Sun Bath.

Hydraulic Motors.—*Water or Hydraulic Motors.* By PHILIP R. BJÖRLING. *With 206 illustrations,* crown 8vo, cloth, 9*s.*

CONTENTS:

1. Introduction—2. Hydraulics relating to Water Motors—3. Water-wheels—4. Breast Water-wheels—5. Overshot and High-breast Water-wheels—6. Pelton Water-wheels—7. General Remarks on Water-wheels—8. Turbines—9. Outward-flow Turbines—10. Inward-flow Turbines—11. Mixed-flow Turbines—12. Parallel-flow Turbines—13. Circumferential-flow Turbines—14. Regulation of Turbines—15. Details of Turbines—16. Water-pressure or Hydraulic Engines—17. Reciprocating Water-pressure Engines—18. Rotative Water-pressure Engines—19. Oscillating Water-pressure Engines—20. Rotary Water-pressure Engines—21. General Remarks and Rules for Water-pressure Engines—22. Hydraulic Rams—23. Hydraulic Rams without Air Vessel in Direct Communication with the Drive Pipe—24. Hydraulic Rams with Air Vessel in Direct Communication with the Drive Pipe—25. Hydraulic Pumping Rams—26. Hydraulic Ram Engines—27. Details of Hydraulic Rams—28. Rules, Formulas, and Tables for Hydraulic Rams—29. Measuring Water in a Stream and over a Weir—Index.

Indicator.—*Twenty Years with the Indicator.* By THOMAS PRAY, Jun., C.E., M.E., Member of the American Society of Civil Engineers. *With illustrations,* royal 8vo, cloth, 12*s.* 6*d.*

Indicator.—*A Treatise on the Richards Steam-Engine Indicator and the Development and Application of Force in the Steam-Engine.* By CHARLES T. PORTER. *With illustrations.* Fourth edition, revised and enlarged, 8vo, cloth, 9*s.*

Induction Coils.—*Induction Coils and Coil Making*: a Treatise on the Construction and Working of Shock, Medical and Spark Coils. By F. C. ALLSOP. *With 118 illustrations,* crown 8vo, cloth, 3*s.* 6*d.*

Iron.—*The Mechanical and other Properties of Iron and Steel in connection with their Chemical Composition.* By A. VOSMAER, Engineer. Crown 8vo, cloth, 6*s.*

CONTENTS:

The metallurgical behaviour of Carbon with Iron and Steel, also Manganese—Silicon—Phosphorus—Sulphur—Copper—Chromium—Titanium—Tungsten—Aluminium—Nickel—Cobalt—Arsenic—Analyses of Iron and Steel, &c.

Iron Manufacture.—*Roll-Turning for Sections in Steel and Iron*, working drawings for Rails, Sleepers, Girders, Bulbs, Ties, Angles, &c., also Blooming and Cogging for Plates and Billets. By ADAM SPENCER. Second edition, *with* 78 *large plates*. Illustrations of nearly every class of work in this Industry. 4to, cloth, 1*l.* 10*s.*

Lime and Cement.—*A Manual of Lime and Cement*, their treatment and use in construction. By A. H. HEATH. Crown 8vo, cloth, 6*s.*

Liquid Fuel.—*Liquid Fuel for Mechanical and Industrial Purposes*. Compiled by E. A. BRAYLEY HODGETTS. *With wood engravings*. 8vo, cloth, 5*s.*

Magneto Hand Telephone.—*The Magneto Hand Telephone*. Its construction, fitting-up, and adaptability to every-day use. By NORMAN HUGHES. *Cuts*, 12mo, cloth, 3*s.* 6*d.*

Mechanics.—*The Essential Elements of Practical Mechanics*, based on the principle of work; designed for Engineering Students. By OLIVER BYRNE, formerly Professor of Mathematics, College for Civil Engineers. Fourth edition, *illustrated by numerous wood engravings*, post 8vo, cloth, 7*s.* 6*d.*

Mechanical Engineering.—*Handbook for Mechanical Engineers*. By HENRY ADAMS, Professor of Engineering at the City of London College, Mem. Inst. C.E., Mem. Inst. M.E., &c. Second edition, revised and enlarged. Crown 8vo, cloth, 6*s.*

CONTENTS:

Fundamental Principles of Mechanics—Varieties and Properties of Materials—Strength of Materials and Structures—Pattern Making—Moulding and Founding—Forging, Welding and Riveting—Workshop Tools and General Machinery—Transmission of Power, Friction and Lubrication—Thermodynamics and Steam—Steam Boilers—The Steam Engine—Hydraulic Machinery—Electrical Engineering—Sundry Notes and Tables.

Mechanical Engineering.—*The Mechanician:* a Treatise on the Construction and Manipulation of Tools, for the use and instruction of Young Engineers and Scientific Amateurs, comprising the Arts of Blacksmithing and Forging; the Construction and Manufacture of Hand Tools, and the various Methods of Using and Grinding them; description of Hand and Machine Processes; Turning and Screw Cutting. By CAMERON KNIGHT, Engineer. *Containing* 1147 *illustrations*, and 397 pages of letter-press. Fourth edition, 4to, cloth, 18*s.*

Mechanical Movements.—*The Engineers' Sketch-Book of Mechanical Movements, Devices, Appliances, Contrivances, Details employed in the Design and Construction of Machinery for every purpose.* Collected from numerous Sources and from Actual Work. Classified and Arranged for Reference. *Nearly* 2000 *Illustrations*. By T. W. BARBER, Engineer. Second edition, 8vo, cloth, 7*s.* 6*d.*

Metal Plate Work.—*Metal Plate Work: its Patterns and their Geometry.* Also Notes on Metals and Rules in Mensuration for the use of Tin, Iron, and Zinc Plate-workers, Coppersmiths, Boiler-makers and Plumbers. By C. T. MILLIS, M.I.M.E. Second edition, considerably enlarged. *With numerous illustrations.* Crown 8vo, cloth, 9*s.*

Metrical Tables.—*Metrical Tables.* By Sir G. L. MOLESWORTH, M.I.C.E. 32mo, cloth, 1*s.* 6*d.*

Mill-Gearing.—*A Practical Treatise on Mill-Gearing, Wheels, Shafts, Riggers, etc.;* for the use of Engineers. By THOMAS BOX. Third edition, *with* 11 *plates.* Crown 8vo, cloth, 7*s.* 6*d.*

Mill-Gearing.—*The Practical Millwright and Engineer's Ready Reckoner;* or Tables for finding the diameter and power of cog-wheels, diameter, weight, and power of shafts, diameter and strength of bolts, etc. By THOMAS DIXON. Fourth edition, 12mo, cloth, 3*s.*

Miners' Pocket-Book.—*Miners' Pocket-Book;* a Reference Book for Miners, Mine Surveyors, Geologists, Mineralogists, Millmen, Assayers, Metallurgists, and Metal Merchants all over the world. By C. G. WARNFORD LOCK, author of 'Practical Gold Mining,' 'Mining and Ore-Dressing Machinery,' &c. Fcap. 8vo, roan, gilt edges, 12*s.* 6*d.*

CONTENTS:

Motive Power—Dams and Reservoirs—Transmitting Power—Weights and Measures—Prospecting—Boring—Drilling—Blasting—Explosives—Shaft Sinking—Pumping—Ventilating—Lighting—Coal Cutting—Hauling and Hoisting—Water Softening—Stamp Batteries—Crushing Rolls—Jordan's Centrifugal Process—River Mining—Ore Dressing—Gold, Silver, Copper Smelting—Treatment of Ores—Coal Cleaning—Mine Surveying—British Rocks—Geological Maps—Mineral Veins—Mining Methods—Coal Seams—Minerals—Precious Stones—Metals and Metallic Ores—Metalliferous Minerals—Assaying—Glossary—List of Useful Books—Index, &c., &c., &c.

Mining and Ore-Dressing Machinery.—By C. G. WARNFORD LOCK, Author of 'Practical Gold Mining.' *Numerous illustrations*, super-royal 4to, cloth, 25*s.*

Mining Machinery.—*Mining Machinery:* a Descriptive Treatise on the Machinery, Tools, and other Appliances used in Mining. By G. G. ANDRÉ, F.G.S., Assoc. Inst. C.E., Mem. of the Society of Engineers. Royal 4to, uniform with the Author's Treatise on Coal Mining, containing 182 *plates*, accurately drawn to scale, with descriptive text, in 2 vols., cloth, 3*l.* 12*s.*

CONTENTS:

Machinery for Prospecting, Excavating, Hauling, and Hoisting—Ventilation—Pumping—Treatment of Mineral Products, including Gold and Silver, Copper, Tin and Lead, Iron, Coal, Sulphur, China Clay, Brick Earth, etc.

Municipal Engineering.—*The Municipal and Sanitary Engineer's Handbook.* By H. PERCY BOULNOIS, Mem. Inst. C.E., Borough Engineer, Portsmouth. *With numerous illustrations.* Second edition, demy 8vo, cloth, 15*s.*

CONTENTS:

The Appointment and Duties of the Town Surveyor—Traffic—Macadamised Roadways—Steam Rolling—Road Metal and Breaking—Pitched Pavements—Asphalte—Wood Pavements—Footpaths—Kerbs and Gutters—Street Naming and Numbering—Street Lighting—Sewerage—Ventilation of Sewers—Disposal of Sewage—House Drainage—Disinfection—Gas and Water Companies, etc., Breaking up Streets—Improvement of Private Streets—Borrowing Powers—Artizans' and Labourers' Dwellings—Public Conveniences—Scavenging, Including Street Cleansing—Watering and the Removing of Snow—Planting Street Trees—Deposit of Plans—Dangerous Buildings—Hoardings—Obstructions—Improving Street Lines—Cellar Openings—Public Pleasure Grounds—Cemeteries—Mortuaries—Cattle and Ordinary Markets—Public Slaughter-houses, etc.—Giving numerous Forms of Notices, Specifications, and General Information upon these and other subjects of great importance to Municipal Engineers and others engaged in Sanitary Work.

Paints.—*Pigments, Paint and Painting.* A Practical Book for Practical Men. By GEORGE TERRY. *With illustrations,* crown 8vo, cloth, 7*s.* 6*d.*

Paper Manufacture.—*A Text-Book of Paper-Making.* By C. F. CROSS and E. J. BEVAN. *With engravings,* crown 8vo, cloth, 12*s.* 6*d.*

Perfumery.—*Perfumes and their Preparation,* containining complete directions for making Handkerchief Perfumes, Smelling Salts, Sachets, Fumigating Pastils, Preparations for the care of the Skin, the Mouth, the Hair, and other Toilet articles, with a detailed description of aromatic substances, their nature, tests of purity, and wholesale manufacture. By G. W. ASKINSON, Dr. Chem. *With* 32 *engravings,* 8vo, cloth, 12*s.* 6*d.*

Perspective.—*Perspective, Explained and Illustrated.* By G. S. CLARKE, Capt. R.E. *With illustrations,* 8vo, cloth, 3*s.* 6*d.*

Petroleum.—*The Marine Transport of Petroleum.* A Book for the use of Shipowners, Shipbuilders, Underwriters, Merchants, Captains and Officers of Petroleum-carrying Vessels. By G. H. LITTLE, Editor of the 'Liverpool Journal of Commerce.' Crown 8vo, cloth, 10*s.* 6*d.*

Phonograph.—*The Phonograph, and How to Construct it.* With a Chapter on Sound. By W. GILLETT. *With engravings and full working drawings,* crown 8vo, cloth, 5*s.*

Pharmacy.—*A Pocket-book for Pharmacists, Medical Practitioners, Students, etc., etc. (British, Colonial, and American).* By THOMAS BAYLEY, Assoc. R. Coll. of Science, Consulting Chemist, Analyst, and Assayer, Author of a 'Pocket-book for Chemists,' 'The Assay and Analysis of Iron and Steel, Iron Ores, and Fuel,' etc., etc. Royal 32mo, boards, gilt edges, 6*s.*

Plumbing.—*Plumbing, Drainage, Water Supply and Hot Water Fitting.* By JOHN SMEATON, C.E., M.S.A., R.P., Examiner to the Worshipful Plumbers' Company. *Numerous engravings*, 8vo, cloth, 7*s.* 6*d.*

Pumping Engines.—*Practical Handbook on Direct-acting Pumping Engine and Steam Pump Construction.* By PHILIP R. BJÖRLING. *With* 20 *plates*, crown 8vo, cloth, 5*s.*

Pumps.—*A Practical Handbook on Pump Construction.* By PHILIP R. BJÖRLING. *Plates*, crown 8vo, cloth, 5*s.*

CONTENTS:

Principle of the action of a Pump—Classification of Pumps—Description of various classes of Pumps—Remarks on designing Pumps—Materials Pumps should be made of for different kinds of Liquids—Description of various classes of Pump-valves—Materials Pump-valves should be made of for different kinds of Liquids—Various Classes of Pump-buckets—On designing Pump-buckets—Various Classes of Pump-pistons—Cup-leathers—Air-vessels—Rules and Formulas, &c., &c.

Pumps.—*Pump Details. With* 278 *illustrations.* By PHILIP R. BJÖRLING, author of a Practical Handbook on Pump Construction. Crown 8vo, cloth, 7*s.* 6*d.*

CONTENTS:

Windbores—Foot-valves and Strainers—Clack-pieces, Bucket-door-pieces, and H-pieces Working-barrels and Plunger-cases—Plungers or Rams—Piston and Plunger, Bucket and Plunger, Buckets and Valves—Pump-rods and Spears, Spear-rod Guides, &c.—Valve-swords, Spindles, and Draw-hooks—Set-offs or Off-sets—Pipes, Pipe-joints, and Pipe-stays—Pump-slings—Guide-rods and Guides, Kites, Yokes, and Connecting-rods—L Bobs, T Bobs, Angle or V Bobs, and Balance-beams, Rock-arms, and Fend-off Beams, Cisterns, and Tanks—Minor Details.

Pumps.—*Pumps and Pumping Machinery.* By F. COLYER, Mem. Inst. C.E., Mem. Inst. M.E. Part I., second edition, revised and enlarged, *with* 50 *plates*, 8vo, cloth, 1*l.* 8*s.*

CONTENTS:

Three-throw Lift and Well Pumps—Tonkin's Patent "Cornish" Steam Pump—Thornewill and Warham's Steam Pump—Water Valves—Water Meters—Centrifugal Pumping Machinery—Airy and Anderson's Spiral Pumps—Blowing Engines—Air Compressors—Horizontal High-pressure Engines—Horizontal Compound Engines—Reidler Engine—Vertical Compound Pumping Engines—Compound Beam Pumping Engines—Shonheyder's Patent Regulator—Cornish Beam Engines—Worthington High-duty Pumping Engine—Davy's Patent Differential Pumping Engine—Tonkin's Patent Pumping Engine—Lancashire Boiler—Babcock and Wilcox Water-tube Boilers.

Pumps.—*Pumps, Historically, Theoretically, and Practically Considered.* By P. R. BJÖRLING. *With 156 illustrations.* Crown 8vo, cloth, 7*s*. 6*d*.

Quantities.—*A Complete Set of Contract Documents for a Country Lodge*, comprising Drawings, Specifications, Dimensions (for quantities), Abstracts, Bill of Quantities, Form of Tender and Contract, with Notes by J. LEANING, printed in facsimile of the original documents, on single sheets fcap., in linen case, 5*s*.

Quantity Surveying.—*Quantity Surveying.* By J. LEANING. With 42 illustrations. Second edition, revised, crown 8vo, cloth, 9*s*.

CONTENTS:

A complete Explanation of the London Practice.
General Instructions.
Order of Taking Off.
Modes of Measurement of the various Trades.
Use and Waste.
Ventilation and Warming.
Credits, with various Examples of Treatment.
Abbreviations.
Squaring the Dimensions.
Abstracting, with Examples in illustration of each Trade.
Billing.
Examples of Preambles to each Trade.
Form for a Bill of Quantities.
Do. Bill of Credits.
Do. Bill for Alternative Estimate.
Restorations and Repairs, and Form of Bill.
Variations before Acceptance of Tender.
Errors in a Builder's Estimate.
Schedule of Prices.
Form of Schedule of Prices.
Analysis of Schedule of Prices.
Adjustment of Accounts.
Form of a Bill of Variations.
Remarks on Specifications.
Prices and Valuation of Work, with Examples and Remarks upon each Trade.
The Law as it affects Quantity Surveyors, with Law Reports.
Taking Off after the Old Method.
Northern Practice.
The General Statement of the Methods recommended by the Manchester Society of Architects for taking Quantities.
Examples of Collections.
Examples of "Taking Off" in each Trade.
Remarks on the Past and Present Methods of Estimating.

Railway Curves.—*Tables for Setting out Curves for Railways, Canals, Roads, etc.*, varying from a radius of five chains to three miles. By A. KENNEDY and R. W. HACKWOOD. *Illustrated*, 32mo, cloth, 2*s*. 6*d*.

Roads.—*The Maintenance of Macadamised Roads.* By T. CODRINGTON, M.I.C.E., F.G.S., General Superintendent of County Roads for South Wales. Second edition, 8vo, cloth, 7*s*. 6*d*.

Safety Valve.—*Safety Valves: their history, antecedents, invention, and calculation*; including the most recent examples of Weighted and Spring-loaded Valves, also showing the effect of Atmospheric Pressure on Safety Valve Discs, showing the curious phenomenon of Balls being sustained by an inclined current of Air; Vacuum Valves, and their importance in heating and boiling. By W. B. LE VAN. *With 69 engravings*, fcap. 8vo, cloth, 6*s*. 6*d*.

Scamping Tricks.—*Scamping Tricks and Odd Knowledge occasionally practised upon Public Works*, chronicled from the confessions of some old Practitioners. By JOHN NEWMAN, Assoc. M. Inst. C.E., author of 'Earthwork Slips and Subsidences upon Public Works,' 'Notes on Concrete,' &c. Crown 8vo, cloth, 2*s*. 6d.

Screw Cutting.—*Turners' Handbook on Screw Cutting, Coning, etc., etc.*, with Tables, Examples, Gauges, and Formulæ. By WALTER PRICE. Fcap. 8vo, cloth, 1*s*.

Screw Cutting.—*Screw Cutting Tables for Engineers and Machinists*, giving the values of the different trains of Wheels required to produce Screws of any pitch, calculated by Lord LINDSAY, M.P. Oblong, cloth, 2*s*.

Screw Cutting.—*Screw Cutting Tables*, for the use of Mechanical Engineers, showing the proper arrangement of Wheels for cutting the Threads of Screws of any required pitch, with a Table for making the Universal Gas-pipe Threads and Taps. By W. A. MARTIN, Engineer. Second edition, oblong, cloth, 1*s*.

Slide Valve.—*A Treatise on a Practical Method of Designing Slide-Valve Gears by Simple Geometrical Construction*, based upon the principles enunciated in Euclid's Elements, and comprising the various forms of Plain Slide-Valve and Expansion Gearing; together with Stephenson's, Gooch's, and Allan's Link-Motions, as applied either to reversing or to variable expansion combinations. By EDWARD J. COWLING WELCH, Mem. Inst. M.E. Crown 8vo, cloth, 6*s*.

Steam Boilers.—*Steam Boilers, their Management and Working on land and sea.* By JAMES PEATTIE. *With illustrations*, crown 8vo, cloth, 5*s*.

CONTENTS:

Water Combustion—Incrustation—Priming—Circulation—Fitting—Stiff for Steam—Soot and Scale effects—Feed—Blowing out—Changing Water—Scale Prevention—Expansion of Boilers—Latent Heat—Firing—Banking Fires—Tube stopping—Concentration of Heat—Boiler Repairs—Explosions, &c., &c.

Steam Engine.—*The Steam Engine considered as a Thermodynamic Machine*, a treatise on the Thermodynamic efficiency of Steam Engines, illustrated by Tables, Diagrams, and Examples from Practice. By JAS. H. COTTERILL, M.A., F.R.S., Professor of Applied Mechanics in the Royal Naval College. Second edition, revised and enlarged, 8vo, cloth, 15*s*.

Steam Engine.—*A Practical Treatise on the Steam Engine*, containing Plans and Arrangements of Details for Fixed Steam Engines, with Essays on the Principles involved in Design and Construction. By ARTHUR RIGG, Engineer, Member of the Society of Engineers and of the Royal Institution of Great Britain. Demy 4to, *copiously illustrated with woodcuts and* 103 *plates*, in one Volume. Second edition, cloth, 25*s*.

This work is not, in any sense, an elementary treatise, or history of the steam engine, but is intended to describe examples of Fixed Steam Engines without entering into the wide domain of locomotive or marine practice. To this end illustrations will be given of the most recent arrangements of Horizontal, Vertical, Beam, Pumping, Winding, Portable, Semi-portable, Corliss, Allen, Compound, and other similar Engines, by the most eminent Firms in Great Britain and America. The laws relating to the action and precautions to be observed in the construction of the various details, such as Cylinders, Pistons, Piston-rods, Connecting-rods, Cross-heads, Motion-blocks, Eccentrics, Simple, Expansion, Balanced, and Equilibrium Slide-valves, and Valve-gearing will be minutely dealt with. In this connection will be found articles upon the Velocity of Reciprocating Parts and the Mode of Applying the Indicator. Heat and Expansion of Steam Governors, and the like. It is the writer's desire to draw illustrations from every possible source, and give only those rules that present practice deems correct.

Steam Engine.—*Steam Engine Management*; a Treatise on the Working and Management of Steam Boilers. By F. COLYER, M. Inst. C.E., Mem. Inst. M.E. New edition, 18mo, cloth, 3*s*. 6*d*.

Steam Engine.—*A Treatise on Modern Steam Engines and Boilers*, including Land, Locomotive and Marine Engines and Boilers, for the use of Students. By FREDERICK COLYER, M. Inst. C.E., Mem. Inst. M.E. *With* 36 *plates*. 4to, cloth, 12*s*. 6*d*.

CONTENTS:

1. Introduction—2. Original Engines—3. Boilers—4. High-Pressure Beam Engines—5. Cornish Beam Engines—6. Horizontal Engines—7. Oscillating Engines—8. Vertical High-Pressure Engines—9. Special Engines—10. Portable Engines—11. Locomotive Engines—12. Marine Engines.

Sugar.—*A Handbook for Planters and Refiners*; being a comprehensive Treatise on the Culture of Sugar-yielding Plants, and on the Manufacture, Refining, and Analysis of Cane, Palm, Maple, Melon, Beet, Sorghum, Milk, and Starch Sugars; with copious Statistics of their Production and Commerce, and a chapter on the Distillation of Rum. By C. G. WARNFORD LOCK, F.L.S., &c.; B. E. R. NEWLANDS, F.C.S., F.I.C., Mem. Council Soc. Chemical Industry; and J. A. R. NEWLANDS, F.C.S., F.I.C. *Upwards of* 200 *illustrations and many plates*, 8vo, cloth, 1*l*. 10*s*.

Surveying.—*A Practical Treatise on the Science of Land and Engineering Surveying, Levelling, Estimating Quantities, etc.*, with a general description of the several Instruments required for Surveying, Levelling, Plotting, etc. By H. S. MERRETT. Fourth edition, revised by G. W. USILL, Assoc. Mem. Inst. C.E. 41 *plates, with illustrations and tables*, royal 8vo, cloth, 12*s*. 6*d*.

Tables of Logarithms.—*A B C Five-Figure Logarithms for general use.* By C. J. WOODWARD, B.Sc. Containing Mantissæ of numbers to 10,000. Log. Sines, Tangents, Cotangents, and Cosines to 10" of Arc. Together with full explanations and simple exercises showing use of the tables. 4*s.*

Tables of Squares.—*Barlow's Tables of Squares, Cubes, Square Roots, Cube Roots, Reciprocals of all Integer Numbers up to* 10,000. Post 8vo, cloth, 6*s.*

Telephones. — *Telephones, their Construction and Fitting.* By F. C. ALLSOP. Second edition, revised and enlarged. *With* 210 *illustrations.* Crown 8vo, cloth, 5*s.*

Tobacco Cultivation.—*Tobacco Growing, Curing, and Manufacturing;* a Handbook for Planters in all parts of the world. Edited by C. G. WARNFORD LOCK, F.L.S. *With illustrations.* Crown 8vo, cloth, 7*s.* 6*d.*

Tropical Agriculture.—*Tropical Agriculture:* a Treatise on the Culture, Preparation, Commerce and Consumption of the principal Products of the Vegetable Kingdom. By P. L. SIMMONDS, F.L.S., F.R.C.I. New edition, revised and enlarged, 8vo, cloth, 21*s.*

Turning.—*The Practice of Hand Turning in Wood, Ivory, Shell, etc.*, with Instructions for Turning such Work in Metal as may be required in the Practice of Turning in Wood, Ivory, etc.; also an Appendix on Ornamental Turning. (A book for beginners.) By FRANCIS CAMPIN. Third edition, *with wood engravings*, crown 8vo, cloth, 3*s.* 6*d.*

Valve Gears. — *Treatise on Valve-Gears*, with special consideration of the Link-Motions of Locomotive Engines. By Dr. GUSTAV ZEUNER, Professor of Applied Mechanics at the Confederated Polytechnikum of Zurich. Translated from the Fourth German Edition, by Professor J. F. KLEIN, Lehigh University, Bethlehem, Pa. *Illustrated*, 8vo, cloth, 12*s.* 6*d.*

Varnish.—*The practical Polish and Varnish-Maker;* a Treatise containing 750 practical Receipts and Formulæ for the Manufacture of Polishes, Lacquers, Varnishes, and Japans of all kinds, for workers in Wood and Metal, and directions for using same. By H. C. STANDAGE (Practical Chemist), author of 'The Artist's Manual of Pigments.' Crown 8vo, cloth, 6*s.*

Ventilation.—*Health and Comfort in House Building; or, Ventilation with Warm Air by Self-acting Suction Power.* With Review of the Mode of Calculating the Draught in Hot-air Flues, and with some Actual Experiments by J. DRYSDALE, M.D., and J. W. HAYWARD, M.D. *With plates and woodcuts.* Third edition, with some New Sections, and the whole carefully revised, 8vo, cloth, 7*s*. 6*d*.

Warming and Ventilating. — *A Practical Treatise upon Warming Buildings by Hot Water*, and upon Heat and Heating Appliances in general; with an inquiry respecting Ventilation, the cause and action of Draughts in Chimneys and Flues, and the laws relating to Combustion. By CHARLES HOOD, F.R.S., F.R.A.S., &c. Re-written by FREDERICK DYE. 8vo, cloth, 15*s*.

Watchwork.— *Treatise on Watchwork, Past and Present.* By the Rev. H. L. NELTHROPP, M.A., F.S.A. *With 32 illustrations*, crown 8vo, cloth, 6*s*. 6*d*.

CONTENTS:

Definitions of Words and Terms used in Watchwork—Tools—Time—Historical Summary—On Calculations of the Numbers for Wheels and Pinions; their Proportional Sizes, Trains, etc.—Of Dial Wheels, or Motion Work—Length of Time of Going without Winding up—The Verge—The Horizontal—The Duplex—The Lever—The Chronometer—Repeating Watches—Keyless Watches—The Pendulum, or Spiral Spring—Compensation—Jewelling of Pivot Holes—Clerkenwell—Fallacies of the Trade—Incapacity of Workmen—How to Choose and Use a Watch, etc.

Waterworks. — *The Principles of Waterworks Engineering.* By J. H. TUDSBERY TURNER, B.Sc., Hunter Medallist of Glasgow University, M. Inst. C.E., and A. W. BRIGHTMORE, M.Sc., Assoc. M. Inst. C.E. *With illustrations*, medium 8vo, cloth, 25*s*.

Well Sinking.—*Well Sinking.* The modern practice of Sinking and Boring Wells, with geological considerations and examples of Wells. By ERNEST SPON, Assoc. Mem. Inst. C.E. Second edition, revised and enlarged. Crown 8vo, cloth, 10*s*. 6*d*.

Wiring. — *Incandescent Wiring Hand-Book.* By F. B. BADT, late 1st Lieut. Royal Prussian Artillery. *With 41 illustrations and 5 tables.* 18mo, cloth, 4*s*. 6*d*.

Wood-working Factories.—*On the Arrangement, Care, and Operation of Wood-working Factories and Machinery*, forming a complete Operator's Handbook. By J. RICHARD, Mechanical Engineer. Second edition, revised, *woodcuts*, crown 8vo, cloth, 5*s*.

SPONS' DICTIONARY OF ENGINEERING,

CIVIL, MECHANICAL, MILITARY, & NAVAL,

WITH

Technical Terms in French, German, Italian, and Spanish.

In 97 numbers, Super-royal 8vo, containing 3132 *printed pages* and 7414 *engravings*. **Any number can be had separate:** Nos. 1 to 95 1*s*. each, post free; Nos. 96, 97, 2*s*., post free.

COMPLETE LIST OF ALL THE SUBJECTS:

Subject	Nos.
Abacus	1
Adhesion	1
Agricultural Engines	1 and 2
Air-Chamber	2
Air-Pump	2
Algebraic Signs	2
Alloy	2
Aluminium	2
Amalgamating Machine	2
Ambulance	2
Anchors	2
Anemometer	2 and 3
Angular Motion	3 and 4
Angle-iron	3
Angle of Friction	3
Animal Charcoal Machine	4
Antimony, 4; Anvil	4
Aqueduct, 4; Arch	4
Archimedean Screw	4
Arming Press	4 and 5
Armour, 5; Arsenic	5
Artesian Well	5
Artillery, 5 and 6; Assaying	6
Atomic Weights	6 and 7
Auger, 7; Axles	7
Balance, 7; Ballast	7
Bank Note Machinery	7
Barn Machinery	7 and 8
Barker's Mill	8
Barometer, 8; Barracks	8
Barrage	8 and 9
Battery	9 and 10
Bell and Bell-hanging	10
Belts and Belting	10 and 11
Bismuth	11
Blast Furnace	11 and 12
Blowing Machine	12
Body Plan	12 and 13
Boilers	13, 14, 15
Bond	15 and 16
Bone Mill	16
Boot-making Machinery	16
Boring and Blasting	16 to 19
Brake	19 and 20
Bread Machine	20
Brewing Apparatus	20 and 21
Brick-making Machines	21
Bridges	21 to 28
Buffer	28
Cables	28 and 29
Cam, 29; Canal	29
Candles	29 and 30
Cement, 30; Chimney	30
Coal Cutting and Washing Machinery	31
Coast Defence	31, 32
Compasses	32
Construction	32 and 33
Cooler, 34; Copper	34
Cork-cutting Machine	34

In demy 4to, handsomely bound in cloth, *illustrated with* 220 *full page plates*, Price 15*s*.

ARCHITECTURAL EXAMPLES

IN BRICK, STONE, WOOD, AND IRON.

A COMPLETE WORK ON THE DETAILS AND ARRANGEMENT OF BUILDING CONSTRUCTION AND DESIGN.

BY WILLIAM FULLERTON, ARCHITECT.

Containing 220 Plates, with numerous Drawings selected from the Architecture of Former and Present Times.

The Details and Designs are Drawn to Scale, 1/8", 1/4", 1/2", and Full size being chiefly used.

The Plates are arranged in Two Parts. The First Part contains Details of Work in the four principal Building materials, the following being a few of the subjects in this Part:—Various forms of Doors and Windows, Wood and Iron Roofs, Half Timber Work, Porches, Towers, Spires, Belfries, Flying Buttresses, Groining, Carving, Church Fittings, Constructive and Ornamental Iron Work, Classic and Gothic Molds and Ornament, Foliation Natural and Conventional, Stained Glass, Coloured Decoration, a Section to Scale of the Great Pyramid, Grecian and Roman Work, Continental and English Gothic, Pile Foundations, Chimney Shafts according to the regulations of the London County Council, Board Schools. The Second Part consists of Drawings of Plans and Elevations of Buildings, arranged under the following heads:—Workmen's Cottages and Dwellings, Cottage Residences and Dwelling Houses, Shops, Factories, Warehouses, Schools, Churches and Chapels, Public Buildings, Hotels and Taverns, and Buildings of a general character.

All the Plates are accompanied with particulars of the Work, with Explanatory Notes and Dimensions of the various parts.

Specimen Pages, reduced from the originals.

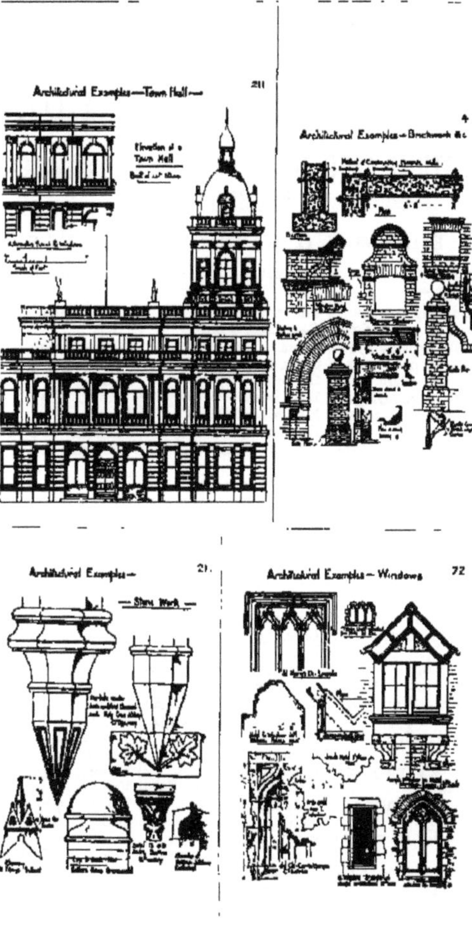

With nearly 1500 *illustrations*, in super-royal 8vo, in 5 Divisions, cloth. Divisions 1 to 4, 13*s*. 6*d*. each; Division 5, 17*s*. 6*d*.; or 2 vols., cloth, £3 10*s*.

SPONS' ENCYCLOPÆDIA

OF THE

INDUSTRIAL ARTS, MANUFACTURES, AND COMMERCIAL PRODUCTS.

EDITED BY C. G. WARNFORD LOCK, F.L.S.

Among the more important of the subjects treated of, are the following:—

Acids, 207 pp. 220 figs.
Alcohol, 23 pp. 16 figs.
Alcoholic Liquors, 13 pp.
Alkalies, 89 pp. 78 figs.
Alloys. Alum.
Asphalt. Assaying.
Beverages, 89 pp. 29 figs.
Blacks.
Bleaching Powder, 15 pp.
Bleaching, 51 pp. 48 figs.
Candles, 18 pp. 9 figs.
Carbon Bisulphide.
Celluloid, 9 pp.
Cements. Clay.
Coal-tar Products, 44 pp. 14 figs.
Cocoa, 8 pp.
Coffee, 32 pp. 13 figs.
Cork, 8 pp. 17 figs.
Cotton Manufactures, 62 pp. 57 figs.
Drugs, 38 pp.
Dyeing and Calico Printing, 28 pp. 9 figs.
Dyestuffs, 16 pp.
Electro-Metallurgy, 13 pp.
Explosives, 22 pp. 33 figs.
Feathers.
Fibrous Substances, 92 pp. 79 figs.
Floor-cloth, 16 pp. 21 figs.
Food Preservation, 8 pp.
Fruit, 8 pp.
Fur, 5 pp.
Gas, Coal, 8 pp.
Gems.
Glass, 45 pp. 77 figs.
Graphite, 7 pp.
Hair, 7 pp.
Hair Manufactures.
Hats, 26 pp. 26 figs.
Honey. Hops.
Horn.
Ice, 10 pp. 14 figs.
Indiarubber Manufactures, 23 pp. 17 figs.
Ink, 17 pp.
Ivory.
Jute Manufactures, 11 pp., 11 figs.
Knitted Fabrics — Hosiery, 15 pp. 13 figs.
Lace, 13 pp. 9 figs.
Leather, 28 pp. 31 figs.
Linen Manufactures, 16 pp. 6 figs.
Manures, 21 pp. 30 figs.
Matches, 17 pp. 38 figs.
Mordants, 13 pp.
Narcotics, 47 pp.
Nuts, 10 pp.
Oils and Fatty Substances, 125 pp.
Paint.
Paper, 26 pp. 23 figs.
Paraffin, 8 pp. 6 figs.
Pearl and Coral, 8 pp.
Perfumes, 10 pp.
Photography, 13 pp. 20 figs.
Pigments, 9 pp. 6 figs.
Pottery, 46 pp. 57 figs.
Printing and Engraving, 20 pp. 8 figs.
Rags.
Resinous and Gummy Substances, 75 pp. 16 figs.
Rope, 16 pp. 17 figs.
Salt, 31 pp. 23 figs.
Silk, 8 pp.
Silk Manufactures, 9 pp. 11 figs.
Skins, 5 pp.
Small Wares, 4 pp.
Soap and Glycerine, 39 pp. 45 figs.
Spices, 16 pp.
Sponge, 5 pp.
Starch, 9 pp. 10 figs.
Sugar, 155 pp. 134 figs.
Sulphur.
Tannin, 18 pp.
Tea, 12 pp.
Timber, 13 pp.
Varnish, 15 pp.
Vinegar, 5 pp.
Wax, 5 pp.
Wool, 2 pp.
Woollen Manufactures, 58 pp. 39 figs.

Crown 8vo, cloth, with illustrations, 5s.

WORKSHOP RECEIPTS,

FIRST SERIES.

SYNOPSIS OF CONTENTS.

Bookbinding.
Bronzes and Bronzing.
Candles.
Cement.
Cleaning.
Colourwashing.
Concretes.
Dipping Acids.
Drawing Office Details.
Drying Oils.
Dynamite.
Electro - Metallurgy — (Cleaning, Dipping, Scratch-brushing, Batteries, Baths, and Deposits of every description).
Enamels.
Engraving on Wood, Copper, Gold, Silver, Steel, and Stone.
Etching and Aqua Tint.
Firework Making — (Rockets, Stars, Rains, Gerbes, Jets, Tourbillons, Candles, Fires, Lances, Lights, Wheels, Fire-balloons, and minor Fireworks).
Fluxes.
Foundry Mixtures.
Freezing.
Fulminates.
Furniture Creams, Oils, Polishes, Lacquers, and Pastes.
Gilding.
Glass Cutting, Cleaning, Frosting, Drilling, Darkening, Bending, Staining, and Painting.
Glass Making.
Glues.
Gold.
Graining.
Gums.
Gun Cotton.
Gunpowder.
Horn Working.
Indiarubber.
Japans, Japanning, and kindred processes.
Lacquers.
Lathing.
Lubricants.
Marble Working.
Matches.
Mortars.
Nitro-Glycerine.
Oils.
Paper.
Paper Hanging.
Painting in Oils, in Water Colours, as well as Fresco, House, Transparency, Sign, and Carriage Painting.
Photography.
Plastering.
Polishes.
Pottery—(Clays, Bodies, Glazes, Colours, Oils, Stains, Fluxes, Enamels, and Lustres).
Scouring.
Silvering.
Soap.
Solders.
Tanning.
Taxidermy.
Tempering Metals.
Treating Horn, Mother-o'-pearl, and like substances.
Varnishes, Manufacture and Use of.
Veneering.
Washing.
Waterproofing.
Welding.

Besides Receipts relating to the lesser Technological matters and processes, such as the manufacture and use of Stencil Plates, Blacking, Crayons, Paste, Putty, Wax, Size, Alloys, Catgut, Tunbridge Ware, Picture Frame and Architectural Mouldings, Compos, Cameos, and others too numerous to mention.

Crown 8vo, cloth, 485 pages, with illustrations, 5s.

WORKSHOP RECEIPTS,

SECOND SERIES.

SYNOPSIS OF CONTENTS

Acidimetry and Alkalimetry.
Albumen.
Alcohol.
Alkaloids.
Baking-powders.
Bitters.
Bleaching.
Boiler Incrustations.
Cements and Lutes.
Cleansing.
Confectionery.
Copying.
Disinfectants.
Dyeing, Staining, and Colouring.
Essences.
Extracts.
Fireproofing.
Gelatine, Glue, and Size.
Glycerine.
Gut.
Hydrogen peroxide.
Ink.
Iodine.
Iodoform.
Isinglass.
Ivory substitutes.
Leather.
Luminous bodies.
Magnesia.
Matches.
Paper.
Parchment.
Perchloric acid.
Potassium oxalate.
Preserving.

Pigments, Paint, and Painting: embracing the preparation of *Pigments*, including alumina lakes, blacks (animal, bone, Frankfort, ivory, lamp, sight, soot), blues (antimony, Antwerp, cobalt, cæruleum, Egyptian, manganate, Paris, Péligot, Prussian, smalt, ultramarine), browns (bistre, hinau, sepia, sienna, umber, Vandyke), greens (baryta, Brighton, Brunswick, chrome, cobalt, Douglas, emerald, manganese, mitis, mountain, Prussian, sap, Scheele's, Schweinfurth, titanium, verdigris, zinc), reds (Brazilwood lake, carminated lake, carmine, Cassius purple, cobalt pink, cochineal lake, colcothar, Indian red, madder lake, red chalk, red lead, vermilion), whites (alum, baryta, Chinese, lead sulphate, white lead—by American, Dutch, French, German, Kremnitz, and Pattinson processes, precautions in making, and composition of commercial samples—whiting, Wilkinson's white, zinc white), yellows (chrome, gamboge, Naples, orpiment, realgar, yellow lakes); *Paint* (vehicles, testing oils, driers, grinding, storing, applying, priming, drying, filling, coats, brushes, surface, water-colours, removing smell, discoloration; miscellaneous paints—cement paint for carton-pierre, copper paint, gold paint, iron paint, lime paints, silicated paints, steatite paint, transparent paints, tungsten paints, window paint, zinc paints); *Painting* (general instructions, proportions of ingredients, measuring paint work; carriage painting—priming paint, best putty, finishing colour, cause of cracking, mixing the paints, oils, driers, and colours, varnishing, importance of washing vehicles, re-varnishing, how to dry paint; woodwork painting).

WORKSHOP RECEIPTS,

FOURTH SERIES,

DEVOTED MAINLY TO HANDICRAFTS & MECHANICAL SUBJECTS.

250 Illustrations, with Complete Index, and a General Index to the Four Series, 5s.

Waterproofing—rubber goods, cuprammonium processes, miscellaneous preparations.

Packing and Storing articles of delicate odour or colour, of a deliquescent character, liable to ignition, apt to suffer from insects or damp, or easily broken.

Embalming and Preserving anatomical specimens.

Leather Polishes.

Cooling Air and Water, producing low temperatures, making ice, cooling syrups and solutions, and separating salts from liquors by refrigeration.

Pumps and Siphons, embracing every useful contrivance for raising and supplying water on a moderate scale, and moving corrosive, tenacious, and other liquids.

Desiccating—air- and water-ovens, and other appliances for drying natural and artificial products.

Distilling—water, tinctures, extracts, pharmaceutical preparations, essences, perfumes, and alcoholic liquids.

Emulsifying as required by pharmacists and photographers.

Evaporating—saline and other solutions, and liquids demanding special precautions.

Filtering—water, and solutions of various kinds.

Percolating and Macerating.

Electrotyping.

Stereotyping by both plaster and paper processes.

Bookbinding in all its details.

Straw Plaiting and the fabrication of baskets, matting, etc.

Musical Instruments—the preservation, tuning, and repair of pianos, harmoniums, musical boxes, etc.

Clock and Watch Mending—adapted for intelligent amateurs.

Photography—recent development in rapid processes, handy apparatus, numerous recipes for sensitizing and developing solutions, and applications to modern illustrative purposes.

In demy 8vo, cloth, 600 pages and 1420 illustrations, 6s.

SPONS' MECHANICS' OWN BOOK;

A MANUAL FOR HANDICRAFTSMEN AND AMATEURS.

CONTENTS.

Mechanical Drawing—Casting and Founding in Iron, Brass, Bronze, and other Alloys—Forging and Finishing Iron—Sheetmetal Working—Soldering, Brazing, and Burning—Carpentry and Joinery, embracing descriptions of some 400 Woods, over 200 Illustrations of Tools and their uses, Explanations (with Diagrams) of 116 joints and hinges, and Details of Construction of Workshop appliances, rough furniture, Garden and Yard Erections, and House Building—Cabinet-Making and Veneering—Carving and Fretcutting—Upholstery—Painting, Graining, and Marbling—Staining Furniture, Woods, Floors, and Fittings—Gilding, dead and bright, on various grounds—Polishing Marble, Metals, and Wood—Varnishing—Mechanical movements, illustrating contrivances for transmitting motion—Turning in Wood and Metals—Masonry, embracing Stonework, Brickwork, Terracotta and Concrete—Roofing with Thatch, Tiles, Slates, Felt, Zinc, &c.—Glazing with and without putty, and lead glazing—Plastering and Whitewashing—Paper-hanging—Gas-fitting—Bell-hanging, ordinary and electric Systems—Lighting—Warming—Ventilating—Roads, Pavements, and Bridges—Hedges, Ditches, and Drains—Water Supply and Sanitation—Hints on House Construction suited to new countries.

E. & F. N. SPON, 125 Strand, London.
New York: 12 Cortlandt Street.

LONDON: PRINTED BY WILLIAM CLOWES AND SONS, LIMITED, STAMFORD STREET AND CHARING CROSS.

Lightning Source UK Ltd.
Milton Keynes UK
UKHW02n0637310718
326503UK00003B/117/P